Lab Manual

Introduction to Marine Biology

FOURTH EDITION

Karleskint, George

Turner, Richard

Small, James

Prepared by

Linda Fergusson-Kolmes
Portland Community College

BROOKS/COLE
CENGAGE Learning™

Australia • Brazil • Japan • Korea • Mexico • Singapore • Spain • United Kingdom • United States

For product information and technology assistance, contact us at
**Cengage Learning Customer & Sales Support,
1-800-354-9706**

For permission to use material from this text or product, submit all requests online at **www.cengage.com/permissions**
Further permissions questions can be emailed to
permissionrequest@cengage.com

ISBN-13: 978-1-133-58980-8

ISBN-10: 1-133-58980-4

Brooks/Cole
20 Davis Drive
Belmont, CA 94002-3098
USA

Cengage Learning is a leading provider of customized learning solutions with office locations around the globe, including Singapore, the United Kingdom, Australia, Mexico, Brazil, and Japan. Locate your local office at **www.cengage.com/global**

Cengage Learning products are represented in Canada by Nelson Education, Ltd.

To learn more about Brooks/Cole, visit
www.cengage.com/brookscole

Purchase any of our products at your local college store or at our preferred online store **www.cengagebrain.com**

Printed in the United States of America
2 3 4 5 6   26 25 24 23 22

Table of Contents

Lab 1

The Scientific Method, the Metric System, and Scientific Notation

Objectives

Upon completion of this lab exercise, you should be able to:

1) Distinguish between a **variable (independent and dependent)**, a **control**, and a **replicate** in an experiment.

2) Use the **scientific method** to determine what environmental cues a response in *Volvox* algae colonies.

3) Name the metric units of measurement for **length**, **volume**, **mass**, and **temperature**.

4) List the metric **superunit** and **subunit** prefixes, include the corresponding metric symbol, and express the measurement value using **scientific notation**.

5) Write large and small numbers using **scientific notation**.

6) Use metric conversion factors to convert values between US customary units of measurement and metric units of measurement.

Background: The Scientific Method

The **scientific method** refers to a specific form of problem solving—including techniques for investigation, gathering new knowledge, and/or correcting previous knowledge—and is used by scientists to plan their research and evaluate the results. The scientific method can be classified as a methodological approach, consisting of several distinct steps.

The first step in the scientific method is **observation**. Observation can be as simple as watching what is going on in the environment around you, taking note of patterns or trends. For example, during his travels to the Galápagos Islands in the Pacific Ocean, Charles Darwin observed that the finches spread out across different islands in the archipelago had different beak shapes and sizes, which led him to propose a tentative explanation or a hypothesis.

A **hypothesis** is a statement that uses logic and reason to explain the information from an observation. A good scientific hypothesis leads to predictions and is usually expressed as an if/then statement. An experiment to test a hypothesis must be structured so that it can be either supported or rejected based on **the data that are collected**. Some of the elements of an experiment to think about are:

1. **Variables**, any factor that can be altered and affect the outcome of the experiment, must be carefully monitored. Modifying only one variable at a time (the **experimental variable or independent variable**) helps ensure that you can effectively determine which variable is responsible for the results you observe. The factor(s) that you count or measure to monitor the effect of the independent variables is the **dependent variable**.

2. You must also have a **control** in your experiment. The subjects in the trial where the independent variable was altered is the experimental set. The trial where the independent variable was not altered is the control set. Controls are very important in science because they help an experimenter make sure that other variables did not alter or affect their experimental results and serve as a comparison for the experimental set. For example, to study the effect of the presence of a predator on sea star behavior, two aquaria could be set up. They would have the same amount of water, at the same temperature, at the same salinity with the same number of sea stars. The experimental aquarium would have chemicals from a predator introduced but the control tank would not. The aquaria would be identical in as many ways as possible except for the variable being studied i.e., the presence of a predator.

3. Also important is having enough **replicates** (repetitions of the various treatments) and an adequate **sample size** (number of observations) within your experiment. The experimenter studying the behavior of sea stars in response to the chemical signal of their predators would never make conclusions based on just one sea star tested just one time. They would make sure they tested multiple sea stars (sample size) and ran the experiment a number of times (replicates)

At the end of an experiment, if the experimenter is satisfied with the results and the overall outcome of the experiment, a **conclusion** can be formed and the initial hypothesis can be supported or rejected. It is VERY important to note that a supported hypothesis does not mean that the hypothesis has been proven true—in fact, all science is exhaustively reviewed by experts in the field before a hypothesis is well-supported by further experiments and analysis and becomes a scientific **theory**.

Many people do not realize it, but the scientific method is common throughout human society and in many seemingly unrelated fields of expertise. For example, farmers carefully monitor their fields and crops, and adjust levels and types of fertilizer and pesticide according to the yields they provide, often working closely with agronomists and agricultural scientists. A store manager closely monitors sales figures of retail products, and adjusts things such as type of inventory and amount in stock, based on factors such as time of year and demand, usually working with advertising and market research experts. Even consumers trying to make the wisest purchase may view commercials and advertising, talk to others about the purchase, read testimonials and reviews, and try out samples of the product or service in question.

In all these cases, a single guiding question emerges: How can I maximize my crop yields? How can I best maximize my sales and move my inventory? What is the best product/service to purchase?

Scientific research is also guided by questions, based on resources that are available, the current level of knowledge regarding the research topic, and overall nature of the questions. While there is not just one way to use the scientific method, scientists generally strive to answer the following questions with their research to facilitate organized investigations:

* What is the overall question we are trying to answer?
* What do we know related to this question? (This can lead to a hypothesis.)
* How will we attempt to answer the question? (What is our experimental design?)
* What are our results?
* What conclusions can we draw from our results?
* What is the value of our conclusions? (Can we answer our questions or form other questions based on our conclusions?)

As earlier stated, there is no one scientific method format, and these questions do not have to be followed in any particular order. Some researchers have a conclusion in mind, then attempt to test this conclusion scientifically; while others may think about the value of potential conclusions before thinking about other parts of a potential investigation.

Merely following the scientific method (or the previous questions) does not guarantee good results. Results may be plagued by errors, experimental designs may fail to work properly, and questions may be difficult to answer with a researcher's conclusions. Problems are opportunities to generate new questions.

Live Organisms: Care and Handling

Live organisms are commonly found in laboratory learning environments. However, we must strive to be careful and responsible when using live organisms for education and/or research. When using live organisms, there are several ethical considerations:

• All live organisms used in laboratories should receive the highest standard of care. There are ethics committees concerned with the care and appropriate use of live organisms for education and research. These committees are comprised of both professionals (scientists) and laypeople (non-scientists), such as legal professionals, research experts, and animal rights activists.

- Organisms may have an entirely different level of sensory perception and development than humans, and appropriate care must be taken when handling and caring for these organisms not to disturb them. For example, fishes are highly sensitive to and may be disturbed by what may seem to humans as minor vibrations, such as tapping on a tank.

- Many organisms commonly found in educational laboratories may not be native. It is very important to follow the instructions given by your lab instructor for proper handling to prevent accidental release of foreign species.

Materials

- You will need to bring a calculator and a pencil with you to class this week.
- Culture of *Volvox*
- Plastic pipettes
- Filtered pond, stream, or lake water
- Experimental chambers (flat, long, shallow plastic dishes or troughs)
- Aluminum foil
- Cardboard
- Desk lamps
- Regular (white) incandescent bulbs
- Green, red, and blue tinted incandescent bulbs
- Salt solution (3.5% salinity, approximately seawater)
- Liquid fertilizer

*Please ask your laboratory instructor if you think of any other materials that may be useful in testing your hypotheses!

Exercise 1
Response of Green Algae to Stimuli

Green algae are primary producers, using photosynthesis to produce food which is subsequently transformed into energy. They form the base of food webs in many aquatic and marine ecosystems. *Volvox* is a colonial freshwater algal organism. Its bright green color and spherical shape make it visible to the naked eye.

1. Observe a *Volvox* culture and, considering factors such as distribution and location, describe what you observed:

2. Today, the primary scientific question to address is: **What are the environmental cues (if any) that cause the *Volvox* colonies to "behave" in the ways you observed?** What could possibly cause the colonies to distribute themselves in certain ways or preferentially select certain areas to inhabit? Write down two questions you have about why you think this might be happening. For example, *"Are the behavioral cues of Volvox due to the presence of higher levels of nutrients in certain parts of the dish?"* Then note any additional questions you may have below.

Question #1:

Question #2:

Additional Questions:

Now rewrite your above questions as hypotheses that you could test. For example if the question is, "Are the behavioral cues due to the presence of higher levels of nutrients in certain parts of the dish?" the hypothesis can be written, "If the cues are due to higher nutrient levels, the *Volvox* colonies will migrate to the end of the trough that contains liquid fertilizer (nutrients)." After you have written your questions as hypotheses, list the hypotheses for any additional questions you came up with. Finally, answer the following questions about your experimental procedure:

- How will you perform this experiment?
- What is your independent variable, what is your dependent variable?
- What is your control?

Hypothesis #1:

Hypothesis #2:

Additional Hypotheses:

Experimental Procedure:

3. Time to test your hypothesis! You have been provided with the supplies on the list of required materials. Transfer some *Volvox* organisms from the culture to your experimental chambers by using a plastic pipette.

***Please ask your laboratory instructor if you think of any other materials that may be useful in testing your hypotheses!

4. Describe how you will specifically set up the experiments to test your hypotheses, what happened during the experiment, the overall results, and what conclusion you have reached.

Hypothesis #1

Experimental procedure:

Results:

Conclusion:

Hypothesis #2

Experimental procedure:

Results:

Conclusion:

More Hypotheses

Use this space to describe any additional experiments you perform, including the hypothesis, procedure, results, and conclusions.

Questions

1. Which of your experiments were useful in answering the main scientific question? Why?

2. Was your main hypothesis supported or not supported by your experiments? Why?

3. How did your results compare to those of your lab mates? What factors do you think account for any major differences? Minor differences?

4. How do you suppose scientists can make sure their results are not just due to random chance? How can you be sure you have "true" results?

Background: Metric System

In the United States, we commonly use United States customary units, also known as Standard units for basic measurements such as length, mass, volume, and temperature. However, the majority of the world's countries as well as nearly all published science use the **metric system**, also known as the **International System of Measurement (SI)**.

The basic units in the metric system include the **meter (m)** for length, the **gram** (g) for mass, the **liter (L)** for volume, and temperature measured in degrees **Celsius** (°C). These units are the base units in the metric system, and can be modified by **subunit** and **superunit** prefixes to increase or decrease the value of the base unit.

Compared to United States customary units, which are difficult to convert due to the wide range of multiplicative and divisive factors required (e.g., there are 5,280 feet in a mile and 3 feet in a yard), metric units are easy to convert as they are based on divisions of 10.

Common U.S. to metric conversion factors:

1 inch = 2.54 cm

1 pound = 0.45 kg

1 quart = 0.946 liters

The International System of Measurement uses the Celsius scale rather than the Fahrenheit scale, commonly used in the U.S. Also based on the metric system of 10, water freezes at 0°C and boils at 100°C, as compared to the 32°F freezing point and 212°F boiling point for standard measure.

Conversion between Celsius and Fahrenheit is very straightforward, illustrated by the following formulas:

$$°F = (°C \times 1.8) + 32$$
$$°C = (°F - 32) / 1.8$$

Table 1-1 Commonly Used SuperUnit and Subunit Prefixes

	Symbol	Value	Scientific Notation	Example
SUPERUNITS				
mega-	M	1,000,000	10^6	1 megagram (Mg) = 10^6 g
kilo-	k	1,000	10^3	1 kilometer (km) = 10^3 m
hecto-	h	100	10^2	1 hectogram (hg) = 100 g
deka	da	10	10^1	1 dekaliter (dal) = 10 l
SUBUNITS				
deci-	d	0.1	10^{-1}	1 decigram (dg) = 0.1 g
centi-	c	0.01	10^{-2}	1 centimeter (cm) = 0.01 m
milli-	m	0.001	10^{-3}	1 milliliter (mL) = 0.001 l
micro-	μ	0.000001	10^{-6}	1 micrometer (μm) = 10^{-6} m
nano-	n	0.000000001	10^{-9}	1 nanogram (ng) = 10^{-9} g
pico-	p	0.000000000001	10^{-12}	1 picoliter (pL) = 10^{-12} L

Exercise 2
Metric Conversion Problems

Using what you have learned about conversions, complete the following problems, showing all of your work.

1. The largest marine bacteria can be up to 750 micrometers in diameter. Convert this to millimeters.

2. The deepest spot on earth, Challenger Deep is 11, 020 m below sea level. Convert this to kilometers.

3. The Gulf Stream can travel 160 kilometers per day. Convert this to meters.

4. How many microliters in a milliliter?

5. Many people keep their pet fish in a 5 gallon fish tank. A gallon is equivalent to 4 quarts. Convert 5 gallons to liters.

6. The world's largest animal, the blue whale, can reach lengths of 110 feet and a mass of over 400,000 pounds. How long can a blue whale get in centimeters? How much can a blue whale weigh in megagrams?

7. Krill are tiny shrimp-like crustaceans that serve as the major food source of blue whales. If the average size of a krill individual is 1.5 cm and its average mass is 0.5 decigrams, and a blue whale eats up to 3,600 kilograms of krill in one day, how many krill individuals are consumed daily? How many meters of krill if they were laid end to end?

8. Some deep-sea vents can reach temperatures of 380°C. Convert this to °F.

9. Salt in ocean water allows it to remain liquid at temperatures below freezing, such as 30°F. Convert this temperature to °C.

10. Water in the deep ocean is often in the range of 3-5 °C. Convert this range to °F.

Background: Scientific Notation

As you may have noticed in the previous section, science often deals with numbers that are either very large or very small, and expressing them as a whole number becomes cumbersome. To deal with this problem, the system of **scientific notation** was developed.

Using scientific notation enables us to express all numbers, regardless of size, as a number between 1 and 10 multiplied by a factor of 10. For example, we can express the number 9,340 as 9.34×10^3 and 0.00934 as 9.34×10^{-3}.

When doing calculations with numbers expressed in scientific notation, there are two rules that can help simplify mathematical problems:

Rule #1: When multiplying numbers with exponents, you add the exponents.
Example: $2^4 \times 2^3 = 16 \times 8 = 128$ and for the exponents $4 + 3 = 7$, so the answer is: 2^7 (which is equal to 128).

Rule #2: When dividing numbers with exponents, you subtract the exponents. Example: $2^4 \div 2^3 = 16 \div 8 = 2$ and for the exponents $4 - 3 = 1$, so the answer is 2^1 (which is equal to 2).

Exercise 3
Scientific Notation Problems

Express the following numbers or expressions in scientific notation.

1. $0.000756 = $ _____

2. $9,358,000 = $ _____

3. $892,400,000,000,000,000 = $ _____

4. $17 / 8,052 = $ _____

5. $1/1000 = $ _____

6. $16,000,000,000,000 / 1,000 = $ _____

7. $55 / 2500 = $ _____

Express the following notations numerically.

1. $7.98 \times 10^{-6} = $ _____

2. $1.01 \times 10^{9} = $ _____

3. $9.987 \times 10^{-3} = $ _____

4. $3.83 \times 10^{1} = $ _____

5. $5.6907853 \times 10^{12} = $ _____

6. $8.70 \times 10^{-4} = $ _____

Optional Exercise 4
Experience with Metric Measures

Deepen your understanding of metric measures by working with common objects and the metric system.

Materials

- Meter stick

- Small cm ruler

- 250 mL beaker

- Empty plastic gallon milk jug

- Top loading balance (g)

- Thermometer (°C)

- Variety of sea shells

- Tap water

*Please ask your laboratory instructor if there are additional materials available that you should be aware of

1. Get into groups of 3 to 4 students. Use the meter stick to determine the height of the members of your group:

Group member 1 _____m

Group member 2_____m

Group member 3_____m

Group member 4_____m

2. Choose two of the shells available. Using the small metric ruler, measure the length and the width of each shell.

Shell 1. _____cm long x _____cm wide

Shell 2. _____cm long x _____cm wide

3. Using the 250 mL beaker and available tap water determine how many milliliters of water it takes to fill up the milk jug.

Milk jug _____mL

4. Using the top loading balance determine the mass of each shell in grams.

Shell 1. _____g

Shell 2. _____g

5. Use the thermometer provided and determine the temperature of the air in your laboratory room.   Then fill the 250 mL beaker with tap water and determine the temperature in degrees Celsius.

Room temperature. _____°C

Tap water temperature _____ °C

Glossary of Terms

Celsius temperature scale: Abbreviated °C, a metric temperature scale in which water boils at 100 °C and freezes at 0 °C.

centimeter: (*centum* - hundred) Abbreviated **cm**, a metric measurement of length, 10^{-2} meters.

control: The portion of the experimental population that does not receive the treatment; all variables are being held constant. Results from the experimental groups for each variable are compared with the measurements from the control group for those variables to determine if differences exist as a result of the experimental "treatment."

cubic centimeter: Abbreviated **cc**, a metric measurement of volume based on a cube with 1 cm sides; also equivalent to 1 milliliter (mm).

dependent variable: The variable being measured in some way in an experiment, to quantify the effect of the independent variable.

experiment: A carefully designed investigation meant to test hypotheses and answer scientific questions.

Fahrenheit temperature scale: Abbreviated **°F**, a traditional temperature scale in the United States in which water boils at 212 °F and freezes at 32 °F.

foot: Abbreviated **ft** or designated by ', the basic unit of length in the United States, derived from the average length of the human foot—12 inches.

gram: (*gramme* - a small weight) Abbreviated **g** or **gm**, the basic metric unit of mass; equal to the mass of one cubic centimeter (milliliter) of water.

hypothesis: (*hypo* - under, less; *tithenai* - to put) A conjecture that is based on previous observation(s) and which serves as the basis for further scientific speculation and/or experimentation.

inch: Abbreviated **in** or designated ", a unit of length in use in the United States; equal to 1/36 of a yard, 1/12 of a foot.

independent variable: The factor being altered to study its effect in an experiment.

kilogram: (*chilioi* - thousand; *gramme* - a small weight) Abbreviated **kg**, a metric measure of mass equal to 1000 grams.

kilometer: (*chilioi* - thousand; *metron* - to measure) Abbreviated **km**, a metric measurement of length, 1000 meters.

liter: (*litra* - a measure) Abbreviated **l**, basic metric measure of volume, equal to the volume of one kilogram of water at maximum density.

meter: (*metron* - to measure) Abbreviated **m**, basic metric measure of length, equal to 100 cm or 39.37 inches.

metric system: (*metron* - to measure) Also known as the **International System of Measurement (SI).** A measurement scale based on multiples of the number 10. The basic units of measurement are: length – meter, mass – gram, and volume – liter.

milliliter: (*mille* - thousandth; *litra* - a measure) Abbreviated **mL**, a metric unit of volume, 10^{-3} liters.

millimeter: (*mille* – thousandth; *metron* - to measure) Abbreviated **mm**, a metric unit of length, 10^{-3} meters.

ounce: (*uncial* - twelfth part) Abbreviated **oz**, a unit of mass originally based on the Roman pound and equal to 1/12 of a Roman pound, now used to designate 1/16 of a standard pound.

pound: (*pondus* - weight) Abbreviated **lb**, a unit of mass used in the United States equal to 16 ounces or about 453.6 gm.

replicate: Repetitions of the various "treatments" within a scientific experiment.

sample size: The number of observations in a scientific experiment.

scientific method: A systematic method for examining, recognizing, and formulating scientific questions; designing experiments; collecting data; and evaluating the results of experiments to further refine the understanding of the initial question.

scientific notation: A numerical system that allows expression of any number, regardless of size, as a number between 1 and 10 multiplied by a factor of 10.

subunit: A unit prefix indicating a smaller amount than the base unit (milligram = 1/1000g).

superunit: A unit prefix indicating a larger amount than the base unit (kilogram = 1000g).

variable: A factor in an experiment that can be altered; for example, temperature is a variable if measurements are made at different temperatures, such as 0°C, 10°C, 25°C, etc.

Lab 2
Organization of Invertebrate Communities

Objectives

Upon completion of this lab exercise, you should be able to:

1) Estimate **population size** using a mark-recapture method.

2) Understand and calculate **population densities** of intertidal invertebrate species.

3) Understand and estimate indices of **population dispersion**.

4) Identify some of the organisms of the **intertidal** community.

Background- Describing Community Structure in a Rocky Intertidal

A field trip to a **rocky intertidal** area is a wonderful opportunity to investigate the ecology of a community. Rocky intertidal areas are often very diverse because the habitat is very complex. The rising and falling of the tides result in areas that vary in the amount of time they are exposed to the air; some regions are only exposed for brief periods and some regions are exposed for many hours. The exposure varies seasonally and contributes to gradients along the shoreline of many **abiotic** factors that are important to the range of an organism (e.g., temperature, salinity). The complex environment also allows for many different kinds of **biotic** interactions as well (e.g., competition). Sometimes is possible to observe an interaction directly e.g., predation but some interactions are more subtle. So it is necessary to make many careful measurements in order to begin to understand the structure of an intertidal community.

A **population** is a group of individuals of the same species living in the same area. One of the first things an ecologist might do is estimate the size of the population. This might sound easy if you take an area and count the individuals you see. However, how do you know how good your estimate is? Perhaps some organisms move a lot, and some are cryptic or very good at hiding. **Mark and recapture** is one way to figure out what fraction of the population your sampling method is able to account for. Once you have an idea of the size of the population of organisms you are interested in you can look at the **population density or** the number of individuals per unit of area or volume. Both biotic and abiotic interactions affect the population density. Furthermore, ecological interactions can also affect the **spatial distribution**, or **dispersion**, of the species along the rocky shore (Figure 2-1). If a pattern of nonrandom dispersion is observed then a biologist can ask why. For example in the spring some species of

Rock Whelks come together in groups of 30 or more to spawn. In this lab exercise you will practice estimating the size of a population using a mark-and-recapture method and then look at the spatial distribution of organisms in the field.

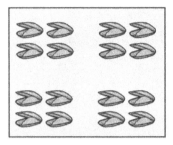  

Figure 2-1 Spatial Distribution. The pattern of spacing of individuals in a population is known as dispersion. Possible patterns are: (a) clumped, (b) uniform, and (c) random.

Materials

- 2 Rectangular plastic containers with lids
- 30 blue marbles
- 200-300 white marbles
- Measuring cup (125 mL or approx ½ cup)
- Hand calculator

Exercise 1
Estimating the Population Size

1. Get into groups of 3 - 4 students. Place 40 blue marbles into the rectangular plastic container you have been given and then add white marbles until approximately two-thirds of the box is full. Place the lid on the plastic container and shake the marbles until they have been thoroughly mixed.

2. Remove the lid from the container and use the measuring cup to scoop out a sample of the marbles.

3. Count the number of blue marbles in your sample and the total number of marbles. Calculate what percentage of your marble sample is blue. E.g., say you have recovered 10 blue marbles out of the 40 blue marbles added. Your sample accounts for about 25% of the blue marbles you know you put in the box. This is your **recapture**. So how many marbles are in the box total? If you recovered 25% of the blue marbles you put in the box then all the marbles in your sample must be 25% of the total. So count the number of blue AND white marbles (say you had 10 blue and 40 white so 50 total). If you have 50 marbles in your sample then there are probably about 200 because if you recovered 50 that is 25% of 200).
Would this really work to estimate a population without counting each individual? Find out.

4. Return the marbles you removed from the box and close the lid. Shake thoroughly. Repeat the sample procedure described two more times to get a total of three samples.

Sample 1. _____blue marbles + _____white marbles=_____total marbles
_____blue marbles ÷ _____total marbles = _____% recapture
Estimate 1 of total number of marbles in box _____ = _____% recapture ÷ _____total marbles

Sample 2. _____blue marbles + _____white marbles=_____total marbles
_____blue marbles ÷ _____total marbles = _____% recapture
Estimate 2 of total number of marbles in box _____ = _____% recapture ÷ _____total marbles

Sample 3. _____blue marbles + _____white marbles=_____total marbles
_____blue marbles ÷ _____total marbles = _____% recapture
Estimate 3 of total number of marbles in box _____ = _____% recapture ÷ _____total marbles

5. What is the **mean** of your three marble population estimates?

The **mean** ($\bar{x}$) is the statistical average of your data set and is calculated by adding observed number of individuals (x_i) in each sample (i) and dividing it by the number of observations (N).

$$\bar{x} = \sum_{i=1}^{N} x_i / N$$

Estimate 1_____ + Estimate 2_____ + Estimate 3_____ = _____ ÷ 3 = _____

6. Now you get to do something you will never be able to do in nature-you can compare your estimate with the real number of marbles in the box. Making sure all marbles have been returned, using the second rectangular box to hold the marbles as you remove them, count the total number of blue and white marbles in your box.

How close was your mean estimate of the marble population to the real population?

What are the limitations of this method in real situations?

Background- Using Statistical Analysis to Describe Dispersion

As explained in the previous exercise certain mathematical methods can be used to evaluate what is observed in nature. For example how do you evaluate if organisms are spatially distributed in clumped, uniform or random patterns? You can look at them and guess, but that leaves room for human error. You will estimate the population of a certain types of rocky intertidal organisms multiple times using a **quadrat**. If the organisms are uniformly dispersed in the environment you expect to see about the same number of the same kind of organism in each sample. If they are clumped then you expect to see more variation. The **mean, variance, standard deviation**, are all used to calculate the **coefficient of dispersion**, which can provide a statistical analysis of the distribution of organisms.

The **mean** ($\bar{x}$) is the statistical average of your data set and is calculated by adding observed number of individuals (x_i) in each quadrat (i) and dividing it by the number of observations (N). You may have noticed in the last exercise that your estimate of the population using the mark-recapture method was not the same each time. How much difference did you see each time? A mathematical way to describe this kind of difference is to calculate the variance between data points.

The **variance** (S^2) is a squared measure of the how spread out the observed data are from the mean. Another way to interpret the variance is the sum of (the difference of each data point to the mean) the quantity squared, in order to have a positive value. The units of the variance are the square of the units of the original data.

$$s^2 = \sum_{i=1}^{n}(\bar{x} - x_i)^2 / (N - 1)$$

The **standard deviation** (S) is the square root of the variance. By taking the square root, the value is linearized, and thus will have the same units as the mean.

The **coefficient of dispersion (CD)** is a measure of the relationship between the variance and the mean of a statistical distribution. Ecologists have learned that for certain statistical distributions, like the **Poisson**, with a random distribution, the mean and the variance are the same. The coefficient of dispersion is calculated by dividing the variance of the numbers of organisms found in your quadrat samples by the mean using the following formula:

$$CD = S^2 / X$$

The coefficient of dispersion is an indicator of the spatial distribution of a species. The approximate values are:

> **CD << 1 – Even distribution**
> **CD ~ 1 – Random distribution**
> **CD >> 1 – Clumped distribution**

Materials

- Measuring tape (at least 20 m)
- 0.5 m^2 quadrat
- Data sheets
- Pencils

Exercise 2
Finding the Distribution Patterns

Figure 2-2 Students using a quadrat

1. Get into groups of three or 3 to 4 students. One student will be in charge of the data recording. The others will count individual organisms within each **quadrat** (Figure 2-2) (each student should count only one species at a time, making sure to turn over rocks, and inspect crevices) return all rocks and organisms to their original positions.

2. Measure a 20 m section parallel to the shore. Try and keep your samples the same distance from the high tide line. Consult with your instructor because the topography of your site will determine what makes sense. (Why would you not chose a constant distance from the water if each group had multiple samples to take?) Starting at the 1-m mark, place the quadrat on the ground. Sort all the mollusk species into one of three major groups (chitons, limpets, and snails) and count how many are present. Repeat every other meter for a total of 10 samples. Record your data on the table below.

3. Calculate the **mean, standard deviation, variance,** and **coefficient of dispersion** for each group and enter the values on the table below.

4. Plot the abundances among the listed organisms on the following graphs:
   - Snails vs. chitons
   - Snails vs. limpets
   - Limpets vs. chitons

DATE: _____ TIME: _____ WEATHER: _____

Sample	Snail Abundance	Limpet Abundance	Chiton Abundance	Others
1				
2				
3				
4				
5				
6				
7				
8				
9				
10				
Mean=				
Variance=				
SD=				
CD=				

$CD = S^2/X$

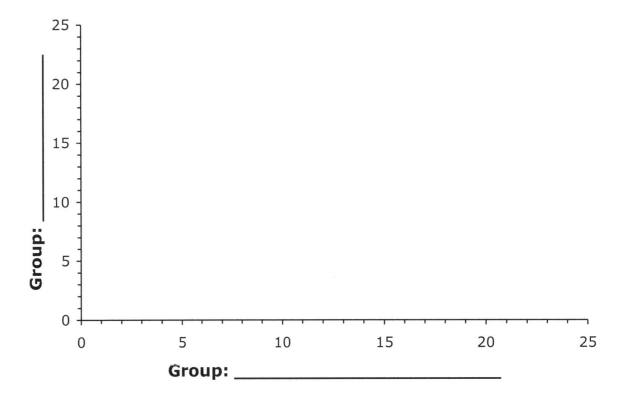

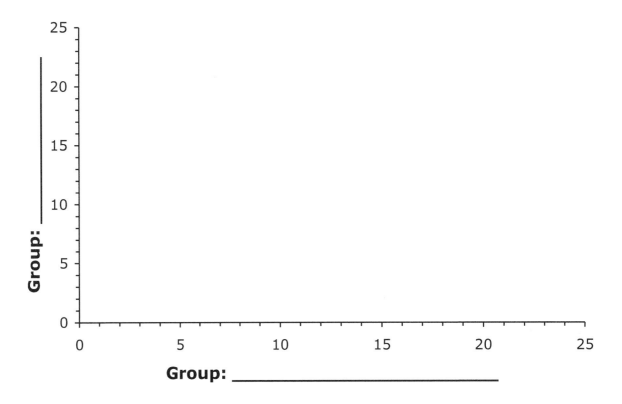

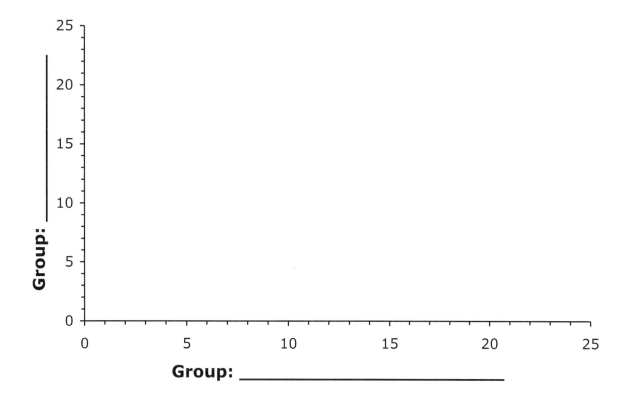

Group: _____

Questions

1. Based on your statistical analysis, what distribution pattern did each species have?

2. What can you infer about the community based on the distributions?

3. Is there a relationship in abundance among any of the groups? Is it positive or negative?

4. What kind of abiotic factors would affect dispersion? Biotic?

5. What effect do you think the size of your quadrat has on the accuracy of your results?

6. What is the effect of the number of quadrat samples taken on the estimate the coefficient of dispersion?

Suggested Readings

Raffaelli, D. F. and S. Hawkins. 1999. Intertidal Ecology. Kluwer Academic Publishers, The Netherlands.

Ricketts, E. and J. Calvin. 1992. Between Pacific Tides. Fifth Edition, Stanford University Press, Stanford, California.

Glossary of Terms

abiotic: Nonliving components in an environment.

biotic: Living components in an environment.

coefficient of dispersion: A statistical measurement that indicates the spatial dispersion of organisms.

competition: Two or more individual organisms of a single species (intraspecific competition), or two or more individuals of different species (interspecific competition), attempting to use the same scarce resources in the same ecosystem.

dispersion: The degree of scatter of organisms.

ecological interaction: The relationships between species that live together in a community.

ecological community: An assembly of populations of different species that occupy the same habitat at the same time.

mark-recapture: A method of estimating population size by marking organism releasing and then recapturing them to estimate the proportion of the total population in a sample.

mean: An average of numbers.

Poisson distribution: A distribution of the expected frequency of events first described by S. D. Poisson early in the 19th century.

population: A group of individuals of the same species that occupies a specified area.

population density: Size of the population within a particular unit of space.

predation: Situation in which an organism of one species (the predator) captures and feeds on parts or all of an organism of another species (the prey).

quadrat: A sampling tool used to mark a specific area of study. It can be made out of PVC piping or string to mark a square meter in the natural environment. For estimating species numbers using pictures, you could use popsicle sticks to mark an area one-tenth the area of the photograph for easier sampling in the classroom.

rocky intertidal: An area of rocky coast that is exposed and submerged by tides.

standard deviation: A measure of how spread out the data is.

variance: The degree of change or difference.

Lab 3

Geocaching with a Global Positioning System

Objectives

Upon completion of this lab exercise, you should be able to:

1) Show a basic knowledge of the **Global Positioning System**.

2) Show familiarity with **GPS receivers**.

3) Show familiarity with the **Universal Transverse Mercator Geographic Coordinate System**.

4) Determine distances among mapped features.

5) Locate field sites when given GPS coordinates.

6) Understand the biological applications for using GPS.

The Global Positioning System

The **Global Positioning System**, or **GPS** as it is commonly known, consists of a network of satellites orbiting some 10,000 miles above the Earth. The purpose of GPS is to allow anyone with a GPS receiver to determine their own position, **latitude** and **longitude**, on Earth within a few meters of error.

Within each satellite there is an atomic clock, which keeps accurate time to the ninth decimal point of a second (that is one billionth of a second). The satellites keep a consistent orbit over the planet, rotating along with the Earth once a day. Therefore, each satellite stays at an exactly constant position over the planet. Each satellite is constantly transmitting a radio signal to Earth, known as a data packet. Each packet consists of the following content: "I am satellite X, my position is Y, and this information was sent at time Z."

The GPS receiver operates by measuring the distance from the satellites that are in orbit around the Earth. The receiver unit does this by estimating the difference in time that each signal had to travel from each respective satellite. By knowing your distance from a number of satellites, it is possible to calculate your position on the Earth's surface by the process of trilateration (analogous to triangulation) Figure 3-1. At least three satellites are required to determine the position of the GPS receiver on the Earth. While this is a great simplification, it is essentially how GPS works.

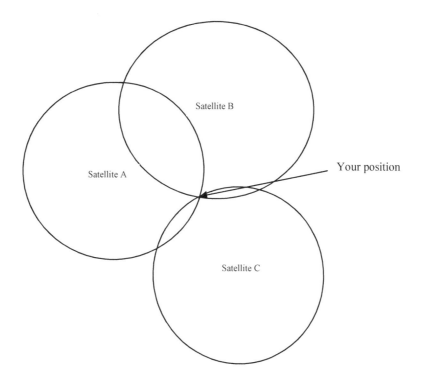

Figure 3-1 Three satellite signals received by a GPS unit would allow the location of the receiver to be determined by trilateration.

GPS Receivers

GPS technology is widely common in society. GPS receivers are found in cars, cell phones, and all sorts of electronic devices. When most people talk about GPS, what usually comes to mind are little handheld devices that look like cell phones. Dedicated GPS receivers have significant advantages over other types when doing scientific work.

GPS receiver models range from less than $100 to professional units costing over $10,000. The two most popular brands are Garmin and Magellan, although several others are available. It is highly recommended that you make sure that if you buy a GPS receiver it is capable of connecting to a computer, usually via a USB (Universal Serial Bus) connection. That way it is easier and faster to download data for future use. Another useful feature is once you have a location fix, to be able to enter text information in the GPS receiver along with the **location record**, to record this as a **waypoint**.

Regardless of their cost, all GPS receivers allow you to determine latitude and longitude of your location, as well as time, and depending on how many satellites you have locked in, even elevation. **Differential Global Positioning Systems** allow a correction can bring the accuracy to within centimeters. However, accuracy of a waypoint can vary based on the receiver quality and atmospheric interference between the receiver and the satellite. Some units will not be able to pick up signals when the user is under inside a building, under heavy foliage or in a slot canyon. Another important feature of most GPS units is that they will provide the location data in many formats (like **UTM**, **Universal Transverse Mercator**), and not just latitude and longitude. This is important in order to determine distances among features that are not located on a map.

Scientists in all areas of marine biology and oceanography use GPS for data collection, navigation, mapping, etc. Researchers can study the behavior of animal species like green sea turtles by attaching specialized GPS units to these animals that travel long distances. These units include sensors that also measure water depth, temperature, salinity, etc., and are equipped with a satellite-triggered transmitter. The units are constantly recording location as well as all other data. At regular intervals, the scientists can contact the GPS unit via satellite and download the data. This allows the scientists to make inferences about turtle behavior, migration patterns, and population dynamics, which until recently would have been impossible. GPS tracking technology can also provide important data about the effectiveness of certain marine conservation measures. Scott et al. (2012) showed that 35% of the world's green sea turtles are congregating in marine protected areas.

Background - Universal Transverse Mercator Geographic Coordinate System

The location of any point on Earth can be described by its latitude (how far north, or south, you are from the equator) and its longitude (how east you are from the **central meridian**) (Figure 3-2).

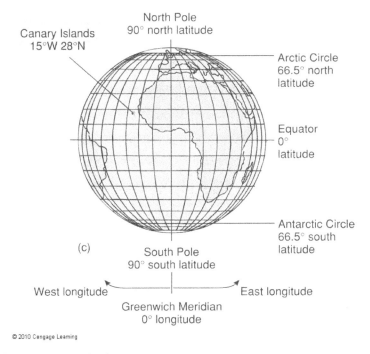

© 2010 Cengage Learning

Figure 3-2 Latitude and Longitude

For precise work latitude and longitude are given in units of degrees, minutes, and seconds. For example, Ambergris Caye, Belize, is located at 17° 59' 36" N latitude and 88° 00' 32" longitude. To calculate the distance in kilometers from Ambergris Caye to where you are located, you would have to use complex mathematical conversions. Alternatively, you could use the **Universal Transverse Mercator** geographic coordinate system, or UTM. The UTM was developed in the 1940's when a series of grid lines at regular intervals were placed on a map based on modified Mercator projections of the Earth (Figure 3-3). The earth is divided into 60 zones, 6 degrees of longitude apart. Oregon, Washington and California would be found in zones 10 and 11 while Florida would fall in UTM zones 16 and 17.

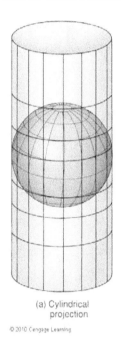

(a) Cylindrical projection

© 2010 Cengage Learning

Using the UTM system, any point can be described by its distance east of the origin (its "easting" value) and its distance north (its "northing" value). By definition, the central meridian is assigned a false easting of 500,000 meters. Any easting value greater than 500,000 meters indicates a point east of the central meridian. Any easting value less than 500,000 meters indicates a point west of the central meridian.

Distances (and locations) in the UTM system are measured in meters, and each UTM zone has its own origin for east-west measurements. Many topographic maps published in recent years use the UTM coordinate system as the primary grid on the map. Because UTM units are in meters it allows you to do the calculations in a direct fashion without the need of conversions.

Figure 3-3 Cylindrical Projection. A Mercator projection is a modification of a cylindrical chart projection

Distance Estimation

It is possible to determine the distance between two features if you have the UTM coordinates. Remember that each tagged feature will have two data points, the easting and the northing, both in meters. This is homologous to coordinates on an *x,y*-plane. If subtracting the northings gives you the distance in meters north to south (side a) and subtracting the eastings gives you the distance in meters east to west (side b) then squaring those values, adding them together and taking the square root will give you the distance between the two features (side c) (Figure 3-4)

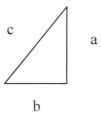

Figure 3-4 Pythagorean theorem $a^2+b^2=c^2$

$$c = \sqrt{(a_2 - a_1)^2 + (b_2 - b_1)^2}$$

To determine the distance between 2 features using UTM coordinates apply equation above where:

c – distance (in meters)

a₂ – northing data point of the second feature

a₁ – northing data point of the first feature

b₂ – easting data point of the second feature

b₁ – easting data point of the first feature

Materials

- GPS receivers that will give both latitude and longitude and UTM coordinates
- Map of the area where exercise will take place
- Ruler
- Hand calculator

Exercise 1
Geocaching and Distance Calculations with a GPS

Prior to this laboratory your instructor has geocached 3 items, in this exercise you will use a GPS unit to locate the items based on the coordinates you are given. You will then use both a map and UTM coordinates to figure out the distances between the geocache sites and your starting location. You must be familiar with how to tag, or mark, a point with the unit; how to view the data; and how to view the location data in the receiver in the various coordinate systems (i.e., latitude and longitude vs. UTM).

1. Get into groups of 3 to 4 students. Each group should have a GPS receiver. The instructor will give all the groups the coordinates (in latitude and longitude) for the first geocache.

2. Select a starting point outside the laboratory room, record a waypoint.

3. Use the coordinates to find the geocache. Use the GPS to determine the UTM coordinates of the location. Record data in Table 3-1. Open the geocache to obtain the next set of coordinates.

4. Repeat step 3 until you have located all 3 geocache sites. Return to the lab.

5. Locate the features you have selected on your map. Determine the distances between the waypoints using the ruler and the scale in the map. Record the estimated distances in Table 3-2.

6. After you have finished using the map to determine the distance between the features you chose in the field, calculate the distances among the features using the UTM points and the distance equation provided above.

7. Compare distances derived from the map to distances calculated.

Table 3-1 Locations

Feature or Object	Lat	Long	UTM E	UTM N
Starting point				
Geocache 1 (G1)				
Geocache 2 (G2)				
Geocache 3 (G3)				

Table 3-2 Distance between locations

	GPS Estimate	Map Estimate
distance between starting point and G 1		
distance between starting point and G2		
distance between starting point and G3		
distance between G1 and G2		
distance between G2 and G 3		
distance between G1 and G3		

Questions

1. Which of the two methods would you consider to be more accurate? Why?

2. Are both methods equally accurate at short (< 20m) as well as longer (> 100 m) distances? Why or why not?

3. Did your GPS unit track the distance between the waypoints automatically? Did it track distance traveled as you looked for the geocache or did it track 'as the crow flies'? How did this compare to your calculated distances?

4. Besides location and distances, what other information could you infer using a GPS unit?

5. How could you incorporate that information in the design of marine biology experiments and studies?

Suggested Reading

El-Rabban, A. 2006. Introduction to GPS: The Global Positioning System. Second Ed, Artech House Publishers, Boston.

Harte, L. and B. Levitan. 2007. GPS Quick Course; Technology, Systems and Operation, Althos Publishing, Fuquay Varina, North Carolina.

Scott, R., Hodgson, D. J., Witt, M.J., Coyne, M.S., Adnyana, W., Blumenthal, J. M., Broderick, A. C., Canbolat, A. C., Catry, P., Ciccione, S., Delcroix, E.,Hitipeuw, C., Luschi, P., Pet-Soede, L., Pendoley, K., Richardson, P. B., Rees, A. F. and B. J. Godley. 2012. Global analysis of satellite tracking data shows that adult green turtles are significantly aggregated in Marine Protected Areas. Global Ecology and Biogeography. doi: 10.1111/j.1466-8238.2011.00757.x

Glossary of Terms

central meridian: An imaginary straight line joining the north and south poles of a planet's disk. It is used as a reference for observers making longitude measurements of features on a planet.

Differential Global Positioning Systems: A system that increases the accuracy of GPS by comparing the location calculated by a GPS signal for a site with a known location. Any differences can be used a correction factor.

fix: A position determined by the navigational unit being used.

Geocaching: A treasure hunting game played by people all around the world using GPS coordinates to locate hidden objects.

Global Positioning System (GPS): An electronic system using a network of satellites to indicate on a computerized receiver the position of a vehicle, ship, person, etc.

latitude: Regularly spaced imaginary lines on Earth's surface, running parallel to the equator.

longitude: Regularly spaced imaginary lines on Earth's surface, running north and south and converging at the poles.

Trilateration: A mathematical process that determines the intersection point of multiple spheres. The intersection point is the location that a GPS waypoint provides and the multiple spheres are determined based on the input of multiple satellites. The input provides the distance of each one to the GPS by providing the time that it takes from satellite to receiver.

Universal Transverse Mercator (UTM) geographic coordinate system: A standardized coordinate system based on the metric system and a division of the Earth into 60 six-degree-wide zones.

waypoint: A fix that will be used as a point of reference or a navigational aid. The term is often used interchangeably with fix when using a GPS unit.

Lab 4

Properties of Seawater: Salinity and pH

Objectives

Upon completion of this exercise, you should be able to:

1) Define **salinity** and describe the difference in salinity in different bodies of water.

2) Describe the effect of salinity and temperature on the **density** of water.

3) Describe the effect of density on the **buoyancy**.

4) Interpret a **Plimsoll** line.

5) Describe the sources of **salt** in salt water.

6) Define the following terms: **acid**, **base**, and **salt**.

7) Define **pH** and relate hydrogen ion concentration to the pH scale.

8) Define **buffer** and describe the buffering capacity of the ocean.

9) Experimentally investigate buffering capacity.

Background - Salinity

The amount of **salt** present in water, or **salinity**, varies drastically between fresh and salt water. The range of salinity in the ocean can be between 33 and 38 ppt (or 38 salt molecules per 1000 water molecules), though it typically averages 35 ppt. Salt in the ocean can come from the land over millions of years being washed from rocks to the sea by streams, rivers, or rain. It can also come from hydrothermal vents and underwater volcanoes or from the air as dust particles.

A variety of factors can affect the salinity of water in the ocean, including river run-off, ice formation, evaporation, and rainfall. Freshwater salinity tends to range from 0.065 to 0.30 ppt. The salinity of coastal areas varies much more than the open ocean. For example, in **estuaries** (bodies of water along coastlines where freshwater from rivers flows to meet salt water from the ocean) there can be a significant salinity gradient. Open ocean salinity can vary latitudinally (Figure 4-1). Evaporation is greater than precipitation in the subtropics (20-30 degrees N and 15-25 degrees S), so surface salinity is greater here than areas like the equator where there is more precipitation.

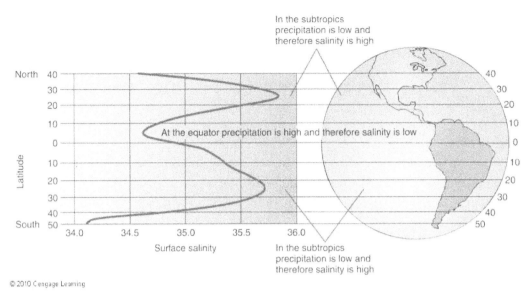

Figure 4-1 Latitudinal Variations in Salinity

Salinity is important to marine organisms for many reasons. The constituents of salt water include some of the important ions that an organism would need to make a shell e.g., Calcium ions, Carbonate ions and silica. Animal cells have selectively permeable membranes and an imbalance between the salt concentration inside a cell and outside a cell can lead to either a loss or gain of water due to **osmosis**. So changes in salinity can present a challenge to homeostasis depending on the organism.

There is an important relationship between the temperature and the salinity of water and its density. Density is the mass per unit volume. A liter of water that had a salinity of 35 ppt would have a greater mass than a liter of fresh water and therefore a greater density. Temperature has an even greater effect on the density of water. The density of water is greatest at about 4°C. Cold salty water sinks. The relative density of water also affects **buoyancy** or the buoyant force. An organism would have an easier time floating in cold, salty water compared to warm freshwater because of the greater upward force exerted by the denser fluid. Organisms regulate their position in the water column in many interesting ways. For example the shelled nautilus is able to rise or sink by controlling the amount of gas present in the chambers inside of its shell. Sharks have livers have a very high concentration of an oil called squalene that helps them maintain their buoyancy. Many bony fish have a gas filled sac called a swim bladder that serves a very similar purpose. Not surprisingly bottom dwelling fish tend not to have a swim bladder.

How does the differences in the density of surface seawater that occur seasonally and geographically affect shipping? Does it matter that a ship will float higher in cold North Atlantic water than in the brackish water of a river port? It turns out that this does matter and that in the

40   Properties of Seawater: Salinity and pH

1860's merchant ships were being lost and sailors drowned because of inappropriate loads. Samuel Plimsoll was instrumental in having a law passed in Britain that lead to load lines being painted on ships. These 'Plimsoll' or loading lines indicated where the water level should be in a ship that was loaded to a safe level. Initially some ship owners resented the law and painted the lines on their funnels. This system of marking has been adopted internationally although there are variations for ship and cargo type. There are different markings for a ship in the summer compared to the winter, seawater compared to fresh and even a mark for tropical water. In this lab you will observe the effect of temperature and salinity on the density of water and subsequent buoyancy. You will use that knowledge to try and figure out how to label a loading line diagram.

Background - The pH Scale

When sodium chloride dissolves in water, it undergoes **dissociation** and separates into **ions** in solution. Water molecules can also dissociate, separating into positively charged hydrogen ions (H^+) and negatively charged hydroxide ions (OH⁻), written as:

$$H_2O \rightarrow H^+ + OH^-$$

The relative concentration of hydrogen or hydroxide ions present can determine whether water is **acidic** or **alkaline (basic)**. When an acid and a base combine, salt and water are formed. The concentration of hydrogen ions in a solution can affect many important aspects of the chemistry of living things. Hydrogen ion concentration is measured using the pH scale. The pH scale ranges from 0 to 14, in which a solution with a pH lower than 7 is acidic; a solution with a pH higher than 7 is basic; and a substance with a pH of 7 (such as pure water at 25 °C) is neutral (Figure 4-2). The pH scale is logarithmic. This means that a change in the pH of a solution from pH 6 to pH 7 would mean a 10 fold drop in hydrogen ion concentration, and a change in the pH of a solution from pH 6 to pH 8 would mean a 100 fold drop in hydrogen ion concentration.

Water found in nature is rarely at pH 7. Many environmental factors can affect the pH of a body of water and the. pH range of freshwater varies widely. The pH of ocean water, remains relatively stable due to buffering. A **buffer** is a substance that can minimize changes in pH by being able to act as a either a base or an acid and keep the relative concentration of hydrogens ions stable. An important buffering system in our blood and in the oceans is the carbonic acid/bicarbonate system. Carbon dioxide is very soluble in water. When carbon dioxide dissolves in water it forms carbonic acid (H_2CO_3). Carbonic acid in turn dissociates into hydrogen ions (H^+) and bicarbonate ions (HCO_3^-). Bicarbonate ions are very important buffers in the ocean because they can act as acids and donate hydrogen ions or as bases and remove hydrogen ions.

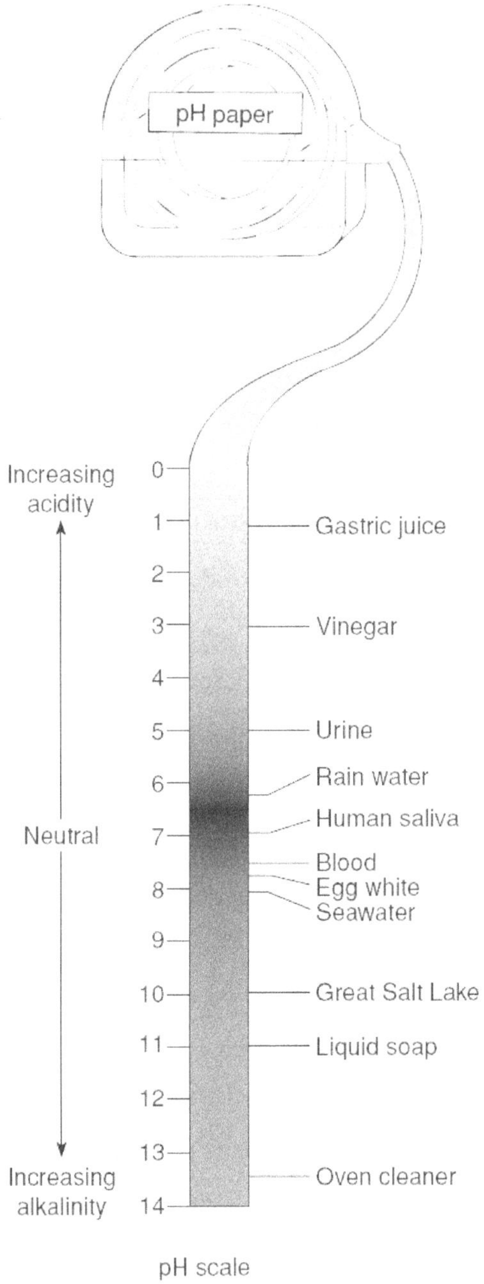

Figure 4-2 pH Scale

pH scale

Even though the ocean's pH is relatively stable, it changes with depth because the amount of carbon dioxide varies with depth. The majority of photosynthetic organisms, such as phytoplankton, live in the top portion of the water column. They use up carbon dioxide during photosynthesis, making the water less acidic. Carbon dioxide is not as soluble in warm water so there is also less carbon dioxide in the photic zone due to the warming of surface waters by sunlight. Due to these factors, the pH in the photic zone tends to average around 8.5. The pH below drops just below the photic zone due to the production of carbon dioxide by marine organisms (such as zooplankton) through respiration. At the bottom of the photosynthetic zone, about 200 m in depth, there is less marine activity, which consequently means less respiration and, therefore, less carbon dioxide.

Materials

- distilled water (or deionized)
- two 50 mL beakers
- labelling tape or a grease pencil
- pens
- 1.0 M HCl (Hydrochloric acid) in a dropper bottle
- 0.1 M HCl (Hydrochloric acid) in a dropper bottle
- pH testing strips (or a pH meter)
- pH color chart (not needed if using a pH meter)
- gloves
- goggles
- 0.1 M sodium acetate solution
- sea water
- freshwater
- cold can of diet soda and cold can of regular soda, identical brand
- NaCl in a 200 mL beaker (¾ full)
- Small aquarium (3-5 gal)
- Refractometer with temperature correction
- waste disposal container for acid

Exercise 1
Observing How the Density of Water Affects Buoyancy

1. Fill the small aquarium with cold tap water until there is at least 5 cm of water on top of a fully submerged soda can sitting on the bottom.

2. Making sure there are no air bubbles on the bottom submerge both the diet soda can and the regular soda can in the aquarium. Push the cans down to the bottom and then release. Note your observations.

3. Shake the NaCl into the aquarium water a little bit at a time, until both soda cans are floating .

4. Use a refractomter to determine the salinity of the water.

5. Empty the aquarium. Repeat steps 1-4 with hot tap water. What differences did you observe?

Exercise 2
Labeling a Loading Line

Below is an unlabelled loading Line (Plimsoll Line). The possibilities for the different kinds of water that a ship might be loaded in are listed:

Tropical Fresh water, Fresh water, Tropical salt water, Saltwater in summer,

Saltwater in winter, and Winter in the North Atlantic.

See if based on what you know about the effect of temperature and salinity on density you can figure out where the markings go. There is a key at the end of this lab.

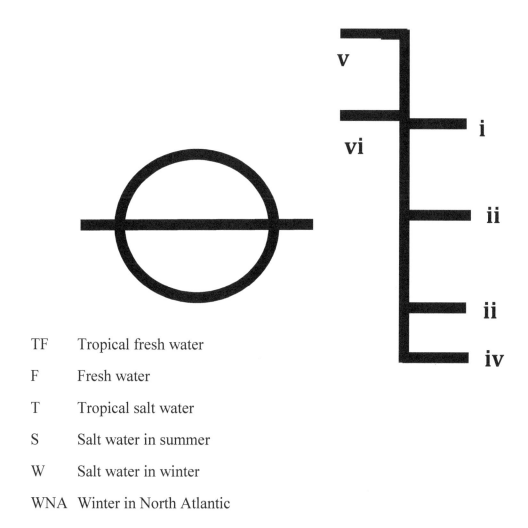

TF   Tropical fresh water

F   Fresh water

T   Tropical salt water

S   Salt water in summer

W   Salt water in winter

WNA   Winter in North Atlantic

Determining pH Change as a Function of Acid Concentration

Safety Note: Be careful when handling acid. You must wear gloves and goggles. If your skin or eyes become exposed to acid, make sure to rinse thoroughly with water and notify your instructor.

1. Place 20 mL of distilled water in two 50 mL beakers. Label one beaker "Concentrated Acid" and the other beaker "Dilute Acid."

2. Measure the pH of the distilled water in each beaker and place your results under the "0" column of Table 1.

3. In the beaker labeled "Concentrated Acid," you will successively add 1–20 drops of 1.0M HCl. Test the pH following the addition of each increment as noted in Table 4-1 using the pH testing strips. Label the strips as you use them so you can keep track of them. Use the pH color chart to find the pH of the solution.

4. Repeat the process for the beaker labeled "Dilute Acid" using 0.1M HCl. Enter your pH results in Table 4-1.

5. Dispose of the contents of the beakers in the waste containers provided.

Table 4-1 Concentrated vs. Dilute Acid in Distilled Water											
# of Acid Drops	0	1	2	3	4	5	10	15	20	25	30
pH of "Concentrated Acid" Beaker											
pH of "Dilute Acid" Beaker											

Exercise 4
Determining pH Change in the Presence of a Buffer

1. Place 20 mL of 0.1 M sodium acetate solution in two 50 mL beakers. Label one beaker "Concentrated Acid" and the other beaker "Dilute Acid."

2. Measure the pH of the sodium acetate solution in each beaker and place your results under the "0" column of Table 4-2.

3. Repeat steps 3-4 from **Experiment 1** and record your data in Table 4-2.

4. Dispose of the contents of the beakers in the waste containers provided.

Table 4-2 Concentrated vs. Dilute Acid in Sodium Acetate Solution											
# of Acid Drops	0	1	2	3	4	5	10	15	20	25	30
pH of "Concentrated Acid" Beaker											
pH of "Dilute Acid" Beaker											

Exercise 5
Natural Buffering Capacity of Seawater

1. Place 20 mL of your seawater sample in two 50 mL beakers. Label one beaker "Concentrated Acid" and the other beaker "Dilute Acid."

2. Measure the pH of the seawater in each beaker and place your results under the "0" column of Table 3.

3. Repeat steps 3-4 from **Experiment 1** and record your data in Table 4-3.

4. Dispose of the contents of the beakers in the waste containers provided

Table 4-3 Concentrated vs. Dilute Acid in Seawater											
# of Acid Drops	0	1	2	3	4	5	10	15	20	25	30
pH of "Concentrated Acid" Beaker											
pH of "Dilute Acid" Beaker											

Exercise 6
Buffering Capacity in Standing Freshwater

1. Place 20 mL of standing freshwater in two 50 mL beakers. Label one beaker "Concentrated Acid" and the other beaker "Dilute Acid."

2. Measure the pH of the standing freshwater in each beaker and place your results under the "0" column of Table 4-4.

3. Repeat steps 3-4 from **Experiment 1** and record your data in Table 4-4.

4. Dispose of the contents of the beakers in the waste containers provided.

Table 4-4 Concentrated vs. Dilute Acid in Standing Freshwater											
# of Acid Drops	0	1	2	3	4	5	10	15	20	25	30
pH of "Concentrated Acid" Beaker											
pH of "Dilute Acid" Beaker											

Plot the data from Tables 4-1to 4-4 on the sheet of graph paper below. Make sure to label both your axes and title your graph.

Reminder: The pH scale is logarithmic!

Questions

1. In the cold tap water did the diet and regular soda cans both float at the same level? Explain your results in terms of density.

2. Did adding the salt to the aquarium water make a difference? If so, why? Did changing the temperature of the aquarium water make a difference? If so, why?

3. Would the loading line on a ship that was only traveling in the Great Lakes be the same as an ocean going ship? Why or why not?

4. Describe the differences in the pH change in the five solutions: the distilled water, the sodium acetate solution, the seawater, and the standing freshwater.

5. Which acid strength (concentrated vs. dilute) had a more significant pH difference when added to solution?

6. Were any of your water samples good buffers? Explain why or why not for each of the samples (seawater and standing freshwater). Consider the source of the water in your answer.

7. Discuss your results of buffering capacity of the seawater sample compared to the artificial buffer.

8. Based on the information you've gathered in the lab, discuss the implication of acid rain and acidic water deposition on the ocean. How might this compare to your two freshwater samples? Include what you have learned about pH, the buffering capacity of seawater and freshwater ecosystems.

9. Upon completion of this exercise, what have you learned about buffering capacity?

Glossary of Terms

acid: A substance that contributes H^+ ions in an aqueous solution.

acidic: Describes a solution with a pH less than 7

base: A substance that removes H^+ ions in an aqueous solution.

basic: Describes a solution with a pH greater than 7, alkaline

brackish: A slightly salt environment, typically with salinity between 2 and 10 ppt.

buffer: A weak acid/base pair that minimizes changes in the pH of a solution .

buffering capacity: The amount of acid or base that can be added to a buffer solution before a significant pH change occurs.

buoyancy: The upward force that keeps an object floating in a liquid.

dissociation: The separation of a molecule into atoms and/or smaller molecules while in solution.

estuary: Bodies of water along coastlines where freshwater from rivers flow to meet salt water from the ocean.

osmosis: The spontaneous movement of a substance in solution from an area of higher concentration to an area of lower concentration across a selectively permeable membrane.

pH: A measure of the amount of hydrogen ions present in a solution, pH = -log $[H^+]$

salinity: The concentration of salt in a solution.

salt: A crystalline structure formed by the replacement of one or more hydrogen ions of an acid when combined with a base in neutralization.

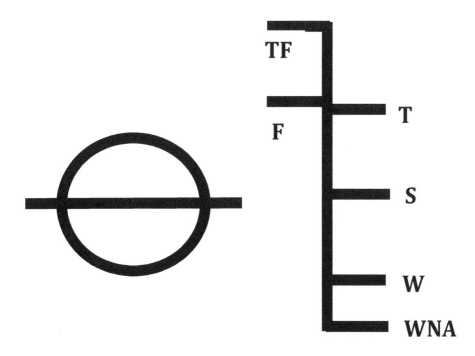

Lab 5
Ocean Acidification

Objectives

Upon completion of this exercise, you should be able to:

1) Describe the **carbon cycle** with respect to a marine environment.

2) Describe the effect of adding **carbon dioxide** to water.

3) Describe different forms of **calcium carbonate** used by marine organisms to build shells (**calcite** and **aragonite**).

4) Experimentally determine the effect of exposing different types of shells to acidic solutions.

5) Discuss the environmental implications of **ocean acidification**.

Background – Ocean Acidification

The amount of carbon dioxide (CO_2) in the earth's atmosphere is increasing due to the burning of fossil fuels and deforestation. Gases can dissolve in water. Carbon dioxide is very soluble in water. Many tons of added CO_2 are removed from the atmosphere by being absorbed by the water in the ocean. Some of the carbon in the water is further trapped by organisms that use calcium carbonate (Ca_2CO_3) to make their shells. The largest reservoir of carbon on the planet is the carbonate system found in marine environments. The ocean acts as a natural sink but the ability of the ocean to absorb CO_2 is not endless and the increase in dissolved CO_2 is lowering the pH of the water. Many scientific studies are being conducted to try to understand how the increased atmospheric CO_2 will affect ocean chemistry and the biology of different organisms.

When carbon dioxide dissolves in water it forms carbonic acid:

$$CO_2 \text{ (aq)} + H_2O \rightleftharpoons H_2CO_3$$

Carbonic acid dissociates into H^+ ions and bicarbonate ions (HCO_3^-) but not completely

$$H_2CO_3 \rightleftharpoons H^+ + HCO_3^-$$

If more H+ ions are added to the system the reaction shifts and more carbonic acid is formed. If H+ ions are removed, more carbonic acid dissociates to replace the H+. This ability to shift back and forth in response to environmental changes is how the carbonate/bicarbonate buffer system works.

Bicarbonate ions can then further dissociate into H+ ions and carbonate ions (CO_3^{2-})

$$HCO_3^- \rightleftharpoons H^+ + CO_3^{2-}$$

The carbonate ions can then be taken up by many organisms and used in the formation of calcium carbonate body parts such as the shells formed by animals such as mussels and clams, the **spicules** in some sponges and the 'skeleton' of some types of corals. The changing pH of the ocean has shifted the availability of carbonate ions. Some studies suggest that a decrease in the available carbonate will slow the growth rate of shell forming organisms.

Studies have also begun to show that some organisms are negatively affected by the decrease in pH because their shells are dissolving. Not all calcium carbonate shells are the same. The crystal structure and the amount of an element called Magnesium (Mg) will affect the relative solubility of a shell. **Aragonite** is type of calcium carbonate shell material that formed by many corals and **pteropods**. Aragonite is more soluble in water than **calcite**, another form of calcium carbonate that has a different crystal structure. Magnesium can be included in the calcite crystal and some forms of calcite are high-magnesium and some are low-magnesium. The more Mg present the more soluble the calcite. The **test** or skeleton of a sea urchin is generally made of low-magnesium calcite. Many organisms make layered shells that alternate between different forms of calcium carbonate e.g., adult oyster shells or many gastropods. Some organisms deposit different kinds of shell depending on the lifestage they are at, e.g., **larval** oysters make their shells from aragonite but mature oysters that have settled make their shells out of much more complex matrix that includes both aragonite and calcite layers. Researchers have documented a link between the increased mortality of the larval Oysters in a commercial hatchery and the CO_2 changes in the ocean off the Oregon coast (Barton et al. 2012). **Paleontologists** have discovered that the chemistry of the ancient oceans has varied and sometimes aragonite shells are favored and sometimes calcite shells are more likely. **Formaniferans** of different species from different time periods have different types of shells. In this lab you will investigate the effect of a dilute acid on different types of shells. You will also add the CO^2 from your breath to water and observe the effect on the pH.

Lab 5   55

Materials for Each Group

- distilled water

- 250 mL beakers (2)

- 100 mL beakers (6)

- Graduated cylinders

- Clean, unused drinking straws

- Bromothymol blue (or Phenol red, a pH indicator)

- 0.4% NaOH (sodium hydroxide) in a dropper bottle

- 5% Acetic Acid (household vinegar)

- Top loading balance (reads to 2 decimal places in grams)

- Hair dryer or drying oven

- Hot plate

- Hot pads for handling glassware

- Forceps

- Mussel shells (or other adult bivalve or gastropod)

- Small pieces of porite coral (clean and dry)

- Paper towels

- Hammer

- Labeling tape or grease pencil

- Gloves

- Goggles

Exercise 1
The Solubility of Different Shell Types

In this exercise you will investigate the effect of exposure to an acidic environment on shells made of different forms of calcium carbonate. You will also investigate the effect of temperature on solubility

1. Get into groups of 3 to 4 students. Turn on the top loading balance and make sure it is reading in grams. Make sure all members of your group are wearing safety goggles and gloves.

2. Select three small pieces of coral. Make sure the coral is clean with no sand particles adhering and is dry. Make sure the pieces are small enough to fit into a 100 mL beaker and be completely covered with 50 mL of liquid. Weigh each piece of coral and record the initial mass in Table 5-1. Place each piece of coral in a separate labeled container.

Table 5-1 Changes in Mass

Item	Initial mass (g)	Final mass (g)	Change (g)	% Change (change in mass ÷ initial mass) x 100
Coral 1				
Coral 2				
Coral 3				
Shell 1				
Shell 2				
Shell 3				

3. Select three small shells. Make sure the shells are clean and dry. If your mussel shells are large, wearing safety goggles, use the hammer to gently break the shell into smaller pieces. Make sure the pieces are small enough to fit into a 100 mL beaker and be completely covered with 50 mL of liquid. Weigh each piece of shell and record the initial mass in Table 5-1. Place each piece of shell in a separate labeled container.

4. Take 6 small beakers (10 mL) and label them: Coral 1, Coral 2, Coral 3, Shell 1, Shell 2 and Shell 3.

5. Measure 50 mL of distilled water into the beakers labeled Coral 1 and Shell 1 Place Coral piece 1 and Shell piece 1 into the water. Record the time and leave in the water for 30 minutes.

6. Measure 50 mL of distilled 5% acetic acid into the beakers labeled Coral 2 and Shell 2 Place Coral piece 2 and Shell piece 2 into the water. Record the time and leave in the vinegar for 30 minutes.

7. Measure another 50 mL of 5% acetic acid into the beakers labeled Coral 3 and Shell 3. Place the beakers on a hot plate and heat until the liquid reaches the boiling point. Using the hot pads remove the beakers from the heat source.
Place Coral piece 3 and Shell piece 3 into the boiling acetic acid. Record the time and leave in the hot acid for 30 minutes.

8. Begin **Exercise 2** while you are waiting.

9. When each shell or coral piece has been soaking in distilled water or 5% acetic acid for 30 minutes, remove with a pair of forceps and blot dry thoroughly with paper towels. When the shells are completely dry, reweigh them. Compare the mass before and after immersion in the liquid (It is critical that the coral and shell pieces are completely dry. If you need to you can use a hair dryer or a drying oven to increase the rate of water evaporation)

10. Calculate % change in mass.

Exercise 2
The Effect of CO_2 on pH

In this exercise you will investigate the effect of CO_2 dissolving in water on pH. You will blow into a beaker that has water and a pH indicator in it. The pH indicator in this experiment is Bromothymol blue. Carbon dioxide is a waste product of respiration. The CO_2 from your exhalation will become H_2CO_3 (carbonic acid). This makes the water acidic, causing the pH indicator to turn a different color. An indirect measure of the amount of CO_2 blown into the water by counting the number of drops of NaOH (a base called sodium hydroxide) needed to return the water to its original color.

1. Get into groups of 3 to 4 students. Measure 200 mL of distilled water into a 250 mL beaker.

2. Add 15 drops of the pH indicator Bromothymol blue to the water in the beaker, until the water is clearly tinted blue. It has a blue color at neutral pH (pH 7). If your water/indicator solution is not a medium blue in color add drops of base until it turns this color.

3. Pour 100 mL of the blue water into the second beaker. The second beaker will serve as a control or color reference.

4. Using a clean drinking straw blow into the test beaker through a straw for 15 seconds. Be sure to wear goggles because the solution can splash up in your face. The solution should turn from blue to yellow as the pH changes and becomes acidic.

5. Add a drop of NaOH (base) to the beaker and swirl gently. Repeat and count the drops, until the solution returns to its original blue color (compare it to your control beaker).

6. Using a clean straw and fresh solutions, have other members of your group repeat steps 1-5.

Questions

1. Did the exposure to the 5% acetic acid affect the coral and the mussel (or other species used) shells equally?

2. What effect did the temperature of the 5% acetic acid have on the outcome?

3. What sources of error are present in your data (e.g., when you put the coral pieces in the liquid did the procedure dislodge any sand)?

4. Many molluscs secrete a proteinaceous layer called a periostracum, as a protective top coat on their shells. Shell collectors often remove this layer by soaking the shell in household bleach. What effect do you think it would have on your results if you removed the periostracum and then exposed the shell to acetic acid?

5. Do you think that all calcium carbonate organisms will be affected equally by the drop in ocean pH? Explain your answer.

6. In exercise 2, based on the number of drops you had to add to neutralize the solution did each member of your group blow the same amount of CO_2 into the water? What variables would affect CO_2 output person-to person?

Suggested Readings

Barton, A., Hales, B., Waldbusser, G.G., Langdon, C. and R. A. Feeley. 2012. The Pacific oyster, Crassostrea gigas, shows negative correlation to naturally elevated carbon dioxide levels: Implications for near-term ocean acidification effects. 57(3) 698–710.

Broeker, W. S. 2003. Treatise on Geochemistry: Chapter 6, The Oceanic CaCO3 Cycle. 6: 529-549. Elsevier

McCulloch, M., Falter, J., Trotter, J., and P. Montagna. 2012. "Coral resilience to ocean acidification and global warming through pH up-regulation." *Nature Climate Change* DOI: 10.1038/nclimate1473

Pacific Marine Environmental Laboratory Ocean Carbon Program

http://www.pmel.noaa.gov/co2/

Glossary of Terms

aragonite: A form of calcium carbonate that can be formed by organisms or by precipitation. It can be distinguished from other forms by its crystal structure.

calcite: A form of calcium carbonate that can be formed by organisms or by precipitation. It can be distinguished from other forms by its crystal structure and is more stable than aragonite.

foraminifera: A diverse group of marine protozoans that form a calcium carbonate test.

larvae: A juvenile form found in many organisms where the adult body form is very different and a metamorphosis must occur.

paleontologist: A scientist who studies fossils.

periostracum: A protective protein layer found on the outside of the shells of mollusks and brachiopods.

pteropods: Sea butterflies, small, pelagic gastropods that swim using wing-like flaps. Some groups form a shell.

spicules: Needle-like structures found in sponges for support, can be made of calcium carbonate or silica.

test: Sometimes refers to the shell-like structure that forms an internal skeleton in for example sea urchins or foraminiferans

Lab 6

Introduction to Microscopes and Cells

Objectives

Upon completion of this exercise, you should be able to:

1) Name and identify the parts of a **compound microscope** and explain their function.

2) Name and identify the parts of a **dissecting microscope** and explain their function.

3) Focus the compound and dissecting microscopes.

4) Prepare a wet mount.

5) Distinguish between a plant cell and an animal cell.

Microscopes and Marine Primary Producers

When you think about the ocean's ecosystem, what do you think of? Dolphins, whales, sharks? Many times our thoughts about something are limited by what we see with our eyes.

However, ecosystems are much more complex than what we see. In fact, most of the primary producers in our ocean's ecosystem cannot be seen with our unaided eye. Primary producers serve as the base of the marine food web, and either directly or indirectly affects all other species, which depend on their productivity in order to survive.

In order to study and understand these organisms we must have specialized tools or instruments. In this lab, we are going to learn how to use a **dissecting microscope** and a **compound microscope**. Dissecting microscopes allow us to see specimens that are unable to fit onto a slide because they are too large. We can observe even smaller specimens with a compound microscope. These microscopes are examples of **light (or optical) microscopes**, which allow us to look at specimens as small as 1 micrometer (μm). To look at specimens even smaller, special microscopes called **electron microscopes** have an electron beam instead of a light beam to illuminate the objects being used.

Materials

- Dissecting microscope
- Penny
- Algae
- Finger bowl or Petri dish
- Compound microscope
- "e" glass slide
- Toothpick
- Stain (e.g., Methylene Blue)
- 0.9% NaCl in dropper bottle
- 2 glass slides
- 2 cover slips
- Paper towels
- *Elodea*
- Scissors

Exercise 1
Using a Dissecting Microscope

In this first exercise you will learn the parts of a dissecting microscope as well as their function.

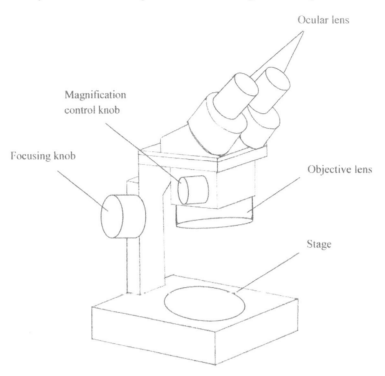

Figure 6-1 Dissecting Microscope

1) Work alone or in pairs, depending on the material available in your lab. Make sure you have the appropriate supplies (dissecting microscope with light source and penny).

2) Place the penny on the **stage**. Adjust the light so that the penny is illuminated sufficiently.

3) Set the **magnification control knob** to the lowest magnification setting possible.

4) The total magnification of the specimen will be the magnification listed on the control knob multiplied by the magnification on the ocular lens.

5) Turn the **focusing knob**, located on the arm of the microscope, until the **objective lens** is as close to the stage as possible.

6) Now, look through the **ocular lenses (eyepiece)**. You should see the penny, but it may not be a clear image.

7) Turn the focusing knob until the image of the penny becomes clear.

8) Now, you will magnify the image on your penny by using the **magnification control knob**. Turn the knob to one magnification setting higher.

9) Examine the magnified penny.

Exercise 2
Dissecting Microscope and Algae

Now that you know how to use a dissecting microscope we can move onto examination of living material.

1) Obtain a Petri dish or finger bowl of algae from your instructor.

2) Make sure that the objective lens of the dissecting microscope is the furthest from the stage as possible. As you recall from Exercise 1, you can adjust this by using the focusing knob.

3) Place the magnification knob on the lowest magnification setting and place the Petri dish with algae on the stage.

4) Adjust the light source so that the algae is sufficiently illuminated. Look through the ocular lenses (eyepiece) to view the algae. Adjust the focus knob.

5) Using the focusing knob, bring the algae into focus.

6) Change the magnification setting two more times and draw what you see in the space provided.

Magnification =	Magnification =	Magnification =

Exercise 3
Compound Microscope

In this first exercise you will learn the parts of a compound microscopes as well as their function.

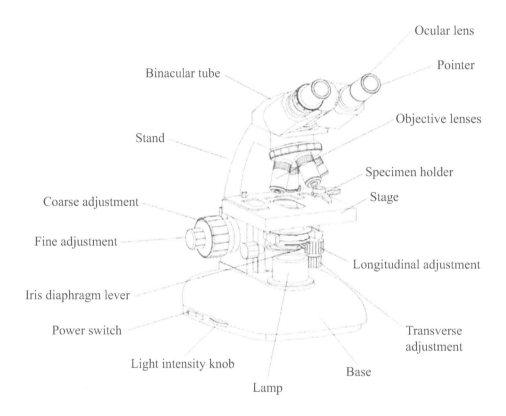

Figure 6-2 Compound Microscope

Structures on a Compound Microscope and their Function

Locate all of the following parts of your microscope. Use Figure 6-2 for reference, although your model may be slightly different.

Ocular Lens (Eyepiece): Magnification value located on upper surface <u>or</u> outer rim
<u>**Caution:**</u> **Do not remove oculars!!** Dirt or dust can enter the system.

Binocular tube: two ocular lenses and tubes = binocular head (microscopes with one ocular = monocular; three oculars = trinocular)

Stand (Arm): carry with one hand <u>**around**</u> this part (other hand under base)

Objective Lens: the magnification value is located around the edge
total magnification = objective value multiplied by ocular value

Note: Objectives with a power **90x or above** are designed to be used **ONLY** with immersion oil and MUST be cleaned thoroughly after using oil.

Power Switch: Turns light source on/off. Turn microscope off if leaving it for more than a few minutes.

Light Intensity Knob: Controls the intensity of the light passing through the slide. Turn to the lowest setting before turning the microscope on/off.

Longitudinal adjustment knob: Controls slide movement from front to back

Transverse adjustment knob: Controls slide movement from side to side
Do NOT move slide with your fingers! This may damage the mechanism that allows the specimen holder to move the slide.

Coarse Adjustment Knob: Used BEFORE fine adjustment to focus the specimen. Turning the knob causes the stage to move up or down.

Fine Adjustment Knob: 'Fine focuses' the image. Turning the knob causes the stage to move up or down, but the movement is very small.

Lamp: Provides a light source.

Base: Carry with other hand **under** base (other around microscope stand).

Iris Diaphragm Lever: Controls amount of light transmitted by opening/closing.

Stage: Flat structure on which microscope slide is placed.

Specimen Holder: Mechanism that holds the slide in place. Pull back the 'hook' on right, insert a slide, and **slowly** release the hook so it does not crack the slide.

Pointer: Right ocular may have a pointer- this allows you to move a slide to line up a specimen or structure with the pointer.

1. Work in alone or in groups depending on the number of microscopes available. Plug the microscope into an electrical outlet and turn the light on, adjust the light intensity so it is the mid-range.

2. If you have a binocular microscope, look through the ocular lenses and adjust the distance until you can see a single circle of light.

3. Obtain a slide with the letter "e" and place the "e" slide on the stage.

4. Look at the slide under low power (4x). Place the slide within the field of view using the transverse and longitudinal adjustment knobs.

5. Use the coarse adjustment knob and put the stage at its lowest point. Slowly roll the stage up using the coarse adjustment knob until you can see the letter "e" on the slide.

6. Adjust the coarse adjustment and fine adjustment knobs until the "e" becomes clearly visible.

7. Change the magnification. Use the fine adjustment knob to sharpen the focus. Do not use the course adjustment knob once you are no longer using the lowest power objective.

8. Draw what you see in the space provided.

9. Determine the total magnification by multiplying the objective magnification by the magnification of the ocular lens (normally the ocular lens is 10x).

10. How does the "e" viewed through the microscope look different from when it is observed without the microscope?

11. Move the slide up and down, side to side. What do you observe? In other words, which way does the "e" appear to move if you move the slide to the right (left, down, up)?

Magnification =	Magnification =	Magnification =

Exercise 4
Compound Microscope and Cheek Cells

In this exercise you will prepare a wet mount of your own cheek cells.

1. Carefully clean a slide. Make sure to hold the slide on the side so that you do not get your fingerprints on the slide.

2. Place a small drop of the sodium chloride (NaCl) solution provided and then a small staining solution (Methylene Blue) on the center of the glass slide. This will stain the cheek cells so that you are able to see specific cellular structures.

3. With the flat end of the toothpick, gently scrape the side of you mouth.

4. Now, swirl the toothpick in the staining solution.

5. Place the cover slip over the cheek cell stain mixture. Make sure that no air bubbles are present.

6. Carefully wipe off the remaining stain that is not covered by the cover slip.

7. Place it under the microscope and draw what you see using three different magnifications. Label any structures that you observe.

8. Use appropriate safety precautions when dealing with human body fluids. Follow the directions given by your lab instructor to dispose of used toothpicks and slides properly.

Magnification =	Magnification =	Magnification =

Exercise 5
Compound Microscope and *Elodea*

In this exercise, you will examine a plant cell using **Elodea**. You will prepare a wet mount using the same technique you used to examine the cheek cell.

1. Prepare a wet mount of the *Elodea* leaf:

    a. First, cut off a small piece of the *Elodea* leaf.

    b. Place a drop of water in the center of the glass slide.

    c. Place the small piece of elodea in the water.

    d. Place the cover slip over the leaf.

    e. Wipe off the excess water outside the cover slip.

2. Place the slide on the microscope stage and examine it using three different magnifications.

3. Draw what you see.

4. Do you notice any particular structures of the plant cell? If so, label them in your drawings.

Magnification =	Magnification =	Magnification =

Questions

1. Name two structures unique to plant cells (not found in animal cells).

2. Name one object you could see using a dissecting (stereo) microscope that you could not observe using a compound microscope.

3. To increase the total magnification from 100x to 400x, describe the process you would use to accomplish this change using names of the various microscope parts involved.

4. If the ocular lens were 10x and the objective were 40x, what would be the total magnification?

Glossary of Terms

Algae: Photosynthetic protists, can be single-celled or multi-cellular. Algae are found in marine and freshwater environments.

cell: The basic unit of all living organisms.

compound microscope: A light (or optical) microscope that has two converging lenses (eyepiece and objective).

dissecting microscope: A stereomicroscope with low-magnifying power; used to view biological specimens or larger objects that do not fit onto a slide

electron microscopes: A microscope that uses electrons rather than visible light to magnify.

Lab 7

Marine Microbes

Objectives

Upon completion of this lab exercise, you should be able to:

1) Identify important characteristics of some representative eubacteria.

2) Understand the complexity associated with classifying **protists**.

3) Describe modes of locomotion in representative protists.

4) Recognize and classify selected members of the protistan **clades** represented in this exercise.

5) Describe the adaptations that help **plankton** remain suspended in the water column.

Eubacteria and Protists

There are many microscopic organisms living in ocean waters. Marine microbes include archeans, bacteria, fungi, protists and many tiny little animals. In this lab we will investigate bacteria and protists. The domain **Eubacteria** contains the prokaryotes known as bacteria. These organisms live in many different habitats and have diverse metabolic capabilities. Domain Eubacteria includes the cyanobacteria, which are photosynthetic **autotrophs**. They make their own food from simple inorganic compounds using light energy to power the process. Cyanobacteria have most of the same photosynthetic pigments that are also found in protistan green algae and members of the kingdom Plantae. They are able to use the sun's energy to split water molecules and use the electrons from the hydrogen atoms to reduce carbon dioxide. This type of photosynthesis is sometimes referred to as modern or oxygenic photosynthesis.

Cyanobacteria were probably the first organisms to have the metabolic machinery to carry out modern photosynthesis. As millions of years passed, they and other modern photosynthetic organisms produced enough oxygen to change the environmental atmosphere and set the stage for the aerobic life forms that were to follow. Other autotrophic bacteria produce their food by **chemosynthesis**, a process that uses energy from inorganic chemical reactions instead of light to produce high-energy organic compounds from low-energy inorganic ones.

The remaining eubacteria are mostly **heterotrophs**—they use food made by other organisms. Many bacteria are important decomposers. In addition to being producers and decomposers, bacteria have other important ecological roles. Some bacteria participate in **nitrogen-fixation**, a process in which atmospheric nitrogen gas is reduced to form ammonia. Some bacteria can oxidize ammonia to nitrites and then to nitrates (**nitrification**). Autotrophs and heterotrophs require a source of either reduced or oxidized nitrogen for producing proteins. Bacteria that participate in nitrogen-fixation and nitrification play an important role in supplying this critical nutrient.

Some bacteria are symbionts of other organisms. There are many examples of symbiosis involving eubacteria in the marine environment. **Symbiosis** is an intimate relationship between members of different species, and there are three main types of symbiosis:

1. Parasitism — one species, the parasite, benefits at the expense of the other, the host.
2. Commensalism — one species benefits and the other is neither harmed nor benefited.
3. Mutualism — both species benefit from the association.

Eubacteria reproduce by binary fission, which results in new individuals that are genetically identical. Genetic change in bacteria is primarily the result of mutation. Mutation does not occur frequently, but bacteria reproduce at a rapid rate, presenting ample opportunity for mutations to arise. This high reproductive capability combined with the sources of genetic variation have produced the great diversity of bacteria found in the marine environment. Some bacteria form thick-walled, dormant spores that are resistant to heat and drying. As a result of this spore-forming ability, the bacteria can survive otherwise lethal environments. When conditions improve, or when the spore is transported to a better habitat, an active cell emerges from the spore.

The domain **Eukarya** contains all the organisms with eukaryotic cells. The cells of eukaryotic organisms contain a nucleus, membrane-bound organelles, and divide by mitosis (or meiosis). The protists are the group of eukaryotic organisms that cannot be classified as fungi, plants, or animals. The protists are found in all types of habitats. Members of this group are metabolically very diverse. They include autotrophs, heterotrophs and **mixotrophs** (organisms that can either make their own food by photosynthesis or can ingest their food).

There are few characteristics that can be applied to all protists. Protists exist in unicellular, filamentous, colonial, and multicellular forms. Some protists are free-living, and others live in close association with other organisms as symbionts (e.g., commensals or parasites). Protists exhibit a vast array of reproductive and feeding mechanisms. Most heterotrophic protists utilize lysosomes to digest food particles in food vacuoles. Gas exchange and the distribution of nutrients and other solutes are accomplished by simple diffusion within the cell and across the cell membrane. Different protists have a variety of structures for locomotion (for example, cilia, flagella, and pseudopods).

For decades, protists have been grouped together into a highly diverse group known as "Kingdom Protista." Kingdom Protista has often been described as a "catch all" kingdom, used to group organisms that didn't fit into Kingdoms Fungi, Plantae, or Animalia. Additionally, the traditional Kingdom Protista was **polyphyletic** (its members descended from two or more ancestral forms that are not common to all members). Recent analyses have shown that many protists are more closely related to organisms in other kingdoms than to each other. For these reasons, and others, the use of "Kingdom Protista" has become obsolete. Protistan systematics is still an area of active research. The Protistan **clades**, shown in Figure 7-1, will be used to organize the representatives surveyed in this lab exercise. In this lab you will observe examples of Eubacteria and some of the major groups of protists.

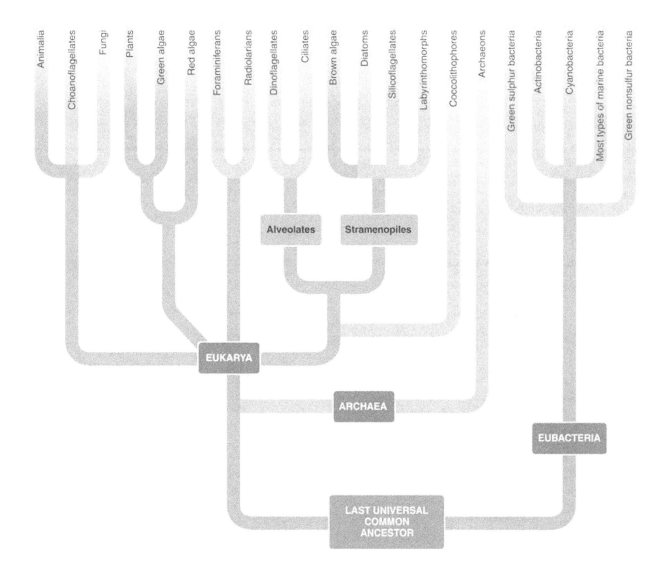

Figure 7-1 Phylogenetic Tree of Life

Materials

- Compound microscope
- Prepared slide of bacterial types (instructor will have prepared as a demo for you to observe)
- *Oscillatoria* culture
- *Anabaena* culture
- Slides
- Coverslips
- Droppers
- Prepared slide of diatoms
- Diatomaceous earth
- Live diatom culture (*Cyclotella* or *Thalassiosira*)
- Live dinoflagellate culture (*Amphidinium* or *Prorocentrum*)
- Live *Paramecium* culture
- Prepared slide of *Ceratium*
- Prepared slide of foraminiferans
- Prepared slide of radiolarians
- Dissecting needle
- Microscope slides and coverslips
- Methyl cellulose
- Container of "build-a-plankton" parts
- Seawater or artificial 'seawater' 30-35 ppt salinity
- 1 L graduated cylinder for floating "plankton" (other containers can be substituted)

Exercise 1
Domain Eubacteria

1. Observe the prepared slide on demonstration that shows the three main shapes of bacteria: **coccus** (spherical), **bacillus** (rod shaped), and **spirillum** (spiral). Use only the fine focus knob to adjust the focus when viewing the slide. Locate at least one bacterial cell of all three shapes and make a sketch of the cells in the space provided. Cell shape is one of the first characteristics used in the classification of bacteria.

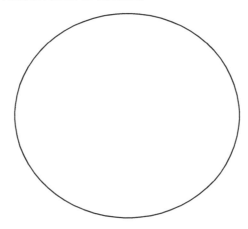

2. Obtain a clean slide and coverslip. Use a dropper to pick up the fine, dark green strands of *Oscillatoria*, a freshwater cyanobacterium, from the culture jar.

3. Place a drop of the culture water on the slide and add a coverslip. Observe *Oscillatoria* under low and high power. The long filaments of *Oscillatoria* are formed by many cells.

4. Draw several filaments of *Oscillatoria* in the space provided. Carefully watch an individual strand of *Oscillatoria* under high power for a few moments. How do you think *Oscillatoria* got its name?

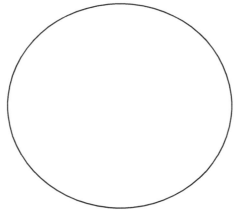

5. Clean your slide carefully and discard the coverslip. Make a wet mount of the *Anabaena* culture.

6. Observe the cells under both low and high power. *Anabaena* cells grow in filaments, here resembling a string of beads. *Anabaena* and other species of cyanobacteria are important ecologically because of their ability to fix nitrogen from the atmosphere into compounds that can be used by algae and plants.

Heterocysts

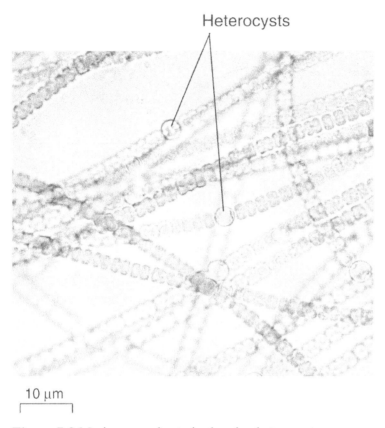

10 µm

Figure 7-2 Marine cyanobacteria showing heterocysts.

7. Note the enlarged cells called **heterocysts (Figure 7-2)** in the *Anabaena* filaments. Nitrogen fixation occurs in these cells. Make a sketch of the cells in the space provided.

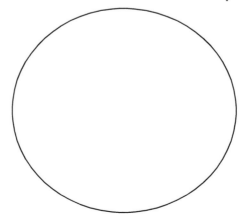

Protists: Stramenopiles

The major groups of marine stramenopiles are diatoms, silicoflagellates, labyrinthomorphs, and brown algae (Phaeophyta). This "hairy flagellum" with the filaments called **mastigonemes** gives this group its name (Figure 7-3). The ecological and economic importance of this group on a global scale is immense. About one-half of all photosynthetic production on Earth comes from algae of different kinds. Not all algae are stramenopiles but many of the major groups are included. Algae are autotrophs, they use the process of photosynthesis to make their own food. All species of algae have the photosynthetic pigment chlorophyll a as well as various accessory pigments. The group to which an alga belons is based in part on the type of accessory pigments, chemical composition of the cell wall and the chemical composition of the stored food.

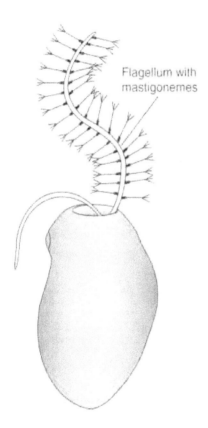

Flagellum with
mastigonemes

Figure 7-3 Stramenopile

Exercise 2
Diatoms

Diatoms (Figure 7-4) are stramenopile members of marine phytoplankton and important primary producers in ocean food chains. They contain chlorophyll *a*, chlorophyll *c*, yellow carotene, and fucoxanthin. The accessory pigments give diatoms a brownish-green or brownish-yellow color. Diatoms are remarkable because of their intricate cell walls, called **frustules**, which consist of two overlapping silica valves. Diatom frustules are used as an abrasive in silver polish and toothpaste and as a filtering material in swimming pools and beer production. When diatom cells die, their siliceous valves accumulate on the ocean floor, producing beds of diatomaceous earth. Diatoms store their food reserves as oil. The oil created by fossilized diatoms is the primary source of the world's oil supply.

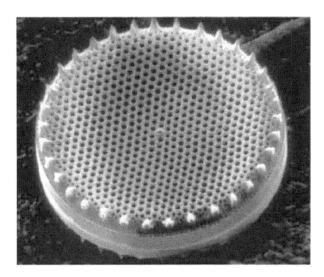

Figure 7-4 A Diatom

1. Using a dissecting needle, scrape a small amount of diatomaceous earth onto a microscope slide and prepare a wet mount slide.

2. Examine your slide, starting with the low magnification objective and proceeding through the highest magnification available on your microscope. Close the iris diaphragm to increase the contrast. You are only looking at the silica frustules. The cytoplasm of this organism disappeared hundreds of thousands, if not millions, of years ago. Note the tiny holes in the frustules.

3. Prepare a wet mount of living diatoms. Use the 40X objective to note the golden brown chloroplasts within the cytoplasm and the numerous holes in the cell walls (frustules). Sketch several of the diatoms you are observing in the space provided.

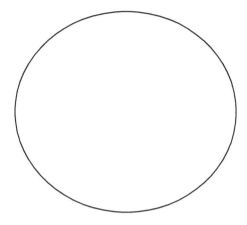

4. Now obtain a prepared slide of diatoms. These cells have been "cleaned," making the perforations in the cell wall especially obvious if you close the iris diaphragm on your microscope's condenser to increase the contrast.

5. Study the frustules with the 40X objective and sketch some examples in the space provided. The pattern of the holes in the walls is characteristic of a given species. Before the advent of electronic techniques, microscopists observed diatom walls to assess the quality of microscope lenses. The resolving power (see discussion of resolving power in the exercise on microscopes) could be determined if one knew the diameter of the holes under observation. Make a sketch of what you are observing in the space provided.

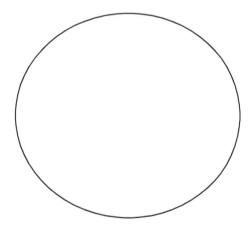

Protists: Alveolates

Alveolates are a group of protists, characterized by membranous sacs (alveoli) beneath their cell membranes. Marine alveolates include dinoflagellates, which are an ecologically important group of phytoplankton, and ciliates, which use dense, hairlike structures known as cilia for locomotion and feeding.

Exercise 3
Dinoflagellates

Dinoflagellates contain chlorophylls *a* and *c* and the accessory pigments beta-carotene and peridinin. Ecologically, dinoflagellates provide a major food source for a diverse array of marine life. Many species have thin cellulose plates that cover their cells, and all dinoflagellates have a characteristic pair of flagella occurring in two grooves within the cell wall, which are used for locomotion (Figure 7-5). Dinoflagellates spin as they move through the water due to the position of their flagella.

During favorable conditions, phytoplankton like dinoflagellates ''bloom'' (undergo rapid population explosions). Some species of dinoflagellate release toxins into the water, causing massive fish die-offs. These events are known as harmful algal blooms (HABs). In some areas, where large populations of toxic dinoflagellates occur, toxins can accumulate in fish and shellfish, making them dangerous to eat. Some species are bioluminescent and, in response to agitation, produce tiny flashes of light in the ocean waters at night.

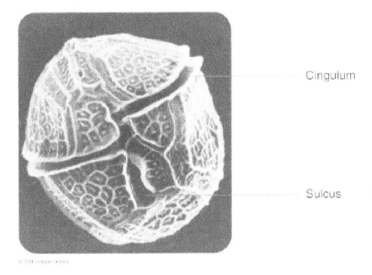

Figure 7-5 An electron micrograph of a dinoflagellate.

1. Using a dropper, place a drop of live dinoflagellate culture on a clean microscope slide and place a coverslip over it. Observe the organisms under low power and high power. You will not be able to see the flagella, but notice the way the organisms move. Adjust the diaphragm and condenser on the microscope to achieve good contrast. Is this a species that has protective cellulose plates? If you look carefully you may be able to see chloroplasts beneath the cell wall. Sketch the organism in the space provided.

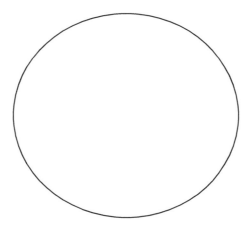

2. Obtain a prepared slide of the dinoflagellate *Ceratium*, and observe it under your compound microscope. Attempt to identify the stiff plates of cellulose encasing the cytoplasm.

3. Locate the two grooves formed by the junction of the plates. This is where the flagella are located.

4. Describe the shape of the dinoflagellate *Ceratium*.

5. Sketch *Ceratium* in the space provided.

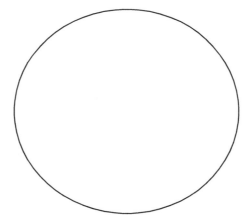

Questions

1. What might be the advantage of the shape of the dinoflagellate *Ceratium*? (Hint: they are potential prey for zooplankton and fish.)

2. What are the characteristics of a eubacterium?

3. What is the first step in the classification of bacteria?

4. The obsolete kingdom Protista was frequently referred to as a "catch all." What does this mean?

5. What are two important roles of cyanobacteria in the marine environment?

6. What characteristic is shared by all members of the Stramenopile clade of protists?

Exercise 4
Ciliates

Ciliates are protozoans that bear cilia for locomotion and gathering food. All ciliates have some portion of the cell surface covered with cilia (singular cilium), which are short, hairlike appendages (Figure 7-6). The cilia beat rapidly in regular, rhythmic patterns that propel the cell through its watery environment. In some species, water currents created by the beating cilia bring food particles to the cell.

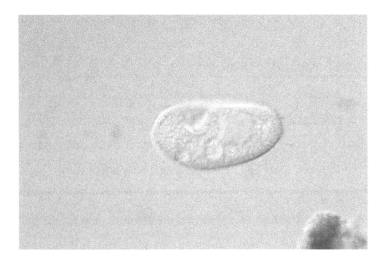

Figure 7-6 *Pleuronema marina* a marine ciliate similar to *Paramecium*

1. Although not a marine organism, we will use *Paramecium* as an example of the protozoans known as ciliates. Obtain a clean slide and coverslip.

2. Look closely at the *Paramecium* culture against a dark background and you may be able to see *Paramecia* as tiny white specks cruising through the water.

3. Use a dropper to draw up some debris from the bottom of the culture jar. Place a few drops of the water onto your slide, then add one drop of methyl cellulose solution. Methyl cellulose is a thick liquid that slows the rapid swimming of *Paramecium* so they can be observed easily under the microscope. Add the coverslip and observe under low and high power.

4. Draw an individual *Paramecium* under high power. Look for the beating cilia by lowering the light of the field and observing the outer edge of the protozoan. Although visible only along the outer boundary of the cell, cilia cover the entire surface of *Paramecium*. Also note the interior of *Paramecium*; look for any organelles.

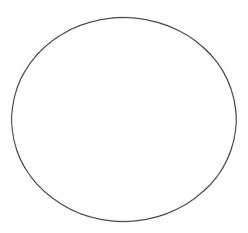

Protists: Foraminiferans and Radiolarians

Foraminiferans are protists that are heterotrophs and so are sometimes referred to as protozoans (animal-like). They have branched **pseudopods** (finger-like extensions of the cytoplasm) that form elaborate netlike structures called **reticulopods.** The reticulopods are used to capture prey, and bottom-dwelling foraminiferans use them to crawl over the bottom sediments. Foraminiferans feed mainly on bacteria and diatoms. The cell of foraminiferans is covered with an elaborate, multi-chambered shell, or **test**, composed of calcium carbonate that frequently resembles a microscopic snail shell (Figure 7-7). Although most foraminiferans are bottom-dwellers, a few species with enormous numbers of individuals are members of the zooplankton. The tests of dead planktonic foraminiferans are a major component of the sediments in the deep ocean.

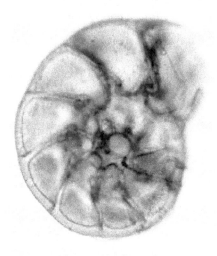

Figure 7-7 Foraminiferans

These sediments are called globigerina ooze because of the large number of tests of the genus *Globigerina*. Over millions of years, geological change has brought some of this sediment to the surface where it forms large deposits of chalk. The White Cliffs of Dover on the English Channel, for example, are formed from globigerina ooze.

Radiolarians have long, needle-like pseudopods called **actinopods**, which function in trapping food (a variety of phytoplankton and zooplankton) and reducing the rate of sinking. The cytoplasm of the cell typically secretes an internal glassy skeleton of silica (Figure 7-8). The skeleton may consist of numerous spines that radiate outward with the pseudopods, resembling a child's jacks. Some species may have an intricate, fused skeleton that resembles a spherical ornament. Since silica is resistant to dissolving in seawater, even under great pressure, the skeletons of dead radiolarians accumulate on the seafloor as radiolarian ooze.

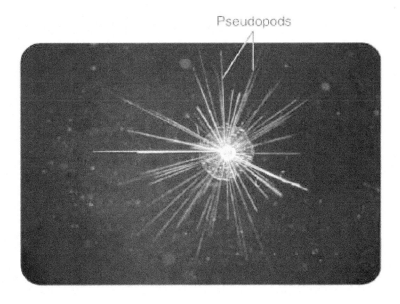

Figure 7-8 A Radiolarian.

Exercise 5
Foraminiferans and Radiolarians

1. Obtain a prepared slide of foraminiferans. The slide contains tests of dead foraminiferans. Examine the slide under low power and high power. Sketch two examples of foraminiferans in the space provided.

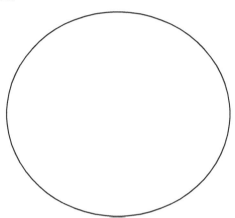

2. Obtain a prepared slide of radiolarians. The slide contains the silica skeletons of dead radiolarians. Examine the slide under low power and high power. Sketch several examples of the skeletons you observe in the space provided.

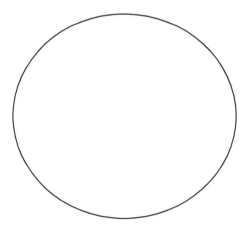

Staying Afloat

Phytoplankton must remain in the sunlit surface waters in order to carry out photosynthesis, and the distribution of phytoplankton dictates the distribution of zooplankton, which that rely on the phytoplankton for food. Since even microbial plankton is denser than seawater, they have to work to keep from sinking out of the sunlit waters.

Some plankton maintain position in the water column by actively swimming. Although swimming slows the rate at which organisms sink, the use of other methods to reduce sinking rates allows them to devote more of their swimming activity to orientation, feeding, and predator avoidance. Adaptations that increase friction (increased surface area) or increase buoyancy (oil droplets, air pockets) help reduce sinking rates and allow marine microbes to devote more energy to other important activities.

Exercise 6
Design Your Own Plankton

1. In this exercise, you will be given access to various materials from which to build your own planktonic organism. Remember: adaptations that will decrease the rate of sinking in the water column will allow the organism to expend more energy on gathering food, avoiding predators, and reproducing.

2. Once you have assembled your plankton, it will be tested by placing it in a 1-liter graduated cylinder filled with seawater.

3. Record the time required for your plankton to sink to the bottom of the cylinder.

4. Compare your results with those of your classmates.

Questions

1. Which group's plankton remained suspended for the longest period of time?

2. What were the characteristics shared by the most successful plankton in the class?

3. List the mode of locomotion utilized by the protists studied in this exercise.

4. How can you distinguish between foraminiferans and radiolarians?

5. Dinoflagellates and ciliates are very different in overall appearance and mode of locomotion, yet they are in the same clade. Why?

6. What are two types of protists that use buoyancy to retard their sinking rate?

7. What radiolarian adaptation helps to reduce their rate of sinking?

Glossary of Terms

actinopods: Needlelike pseudopods that are typical of radiolarians.

autotroph: An organism that is capable of producing its own food from an inorganic source of energy e.g., sunlight.

bacillus: A bacterium with a cell shaped like a rod.

cingulum: A furrow or groove in the outer structure of some dinoflagellates where one of the two flagella can be found, from the Latin for girdle

chemosynthesis: The process of using energy from chemical reactions to build food molecules.

clade: A grouping organisms that reflects their common ancestry

coccus: A bacterium with a cell shaped like a sphere.

Eubacteria: Living things are grouped into three domains. Eubacteria is one of the two domains that contains small, single-celled prokaryotic cells. The other is domain Archaea

Eukarya: The domain containing all the organisms with eukaryotic cells.

frustules: The glassy structures composed of silica that cover diatom cells.

heterocysts: A specialized cell of cyanobacteria in which conditions favorable for nitrogen-fixation are maintained.

heterotroph: An organism that relies on other organisms for food.

mixotroph: an organism that both produce its own food using inorganic forms of energy (autotroph) or consume organic forms of energy (heterotroph).

mastigonemes: One of many hairlike filaments that extend from the shaft of some flagella and that increase the effectiveness of locomotion.

nitrification: The process by which ammonia is converted into nitrate ions.

nitrogen-fixation: The process by which some microorganisms are able to convert atmospheric nitrogen into a form that is useable by autotrophs.

plankton: Organisms that live in the water and are unable to move against currents.

polyphyletic: A term used to indicate that members of a group of organisms descended from two or more ancestral forms that are not common to all of the members.

pseudopods: Fingerlike projection of cytoplasm and membrane that functions in feeding and locomotion in amoeboid protozoans.

reticulopods: Pseudopods with branches that interconnect to form a net for the capture of particles.

spirillum: A bacterium with a cell shaped like a corkscrew.

sulcus: A depression in the outer structure of some dinoflagellates that is the point of insertion for one of the two flagella found in these organisms.

symbiosis: An intimate living arrangement between two different species of organism.

test: Sometimes refers to the shell-like structure that forms as a support structure in foraminiferans, the same term is used to describe the internal skeleton in sea urchins.

Lab 8

Higher Invertebrates: Molluscs

Objectives

Upon completion of this exercise, you should be able to:

1) Recognize representative examples of phylum **Mollusca**.

2) Describe the characteristics of molluscs from the 5 major classes: **Polyplachophora**. **Scaphopoda, Gastropoda, Bivalvia** and **Cephalopoda**.

3) Recognize and describe the function of the structures visible in a representative bivalve.

4) Recognize and describe the function of the visible structures in a representative cephalopod.

Background – Phylum Mollusca

The Molluscs are one of the most successful invertebrate groups, with over 50,000 described living species and a fossil record that dates back to the Cambrian. They are the largest marine phylum and the members of this group are extremely diverse. The generalized mollusc is considered to be a bilaterally symmetrical, benthic organism with a single shell. The main body (visceral mass) is covered and protected by a curtain of protective tissue called the mantle. The mantle secretes the shell. Modern molluscs have radiated to include many pelagic species that lack a shell as well as a vast array of shelled species. The shell is secreted by the mantle of the organism and can be a complex matrix of different kinds of calcium carbonate crystals and protein. The variety of shell shapes, sizes and colors have captivated collectors around the word. They also include many economically important groups. In 2011 the commercial clam industry in Florida was estimated to be worth over 20 million dollars.

There are five major classes of molluscs (and some authorities list an additional two classes the monoplacophorans and aplacophorans that will not be dealt with here). Class **Polyplacophora** are the chitons. These organisms are characterized by having series of 8 shell plates that run up the dorsal side of a flattened body. Chitons are benthic grazers and feed on the organisms they scrape off rocks and other hardened surfaces with their **radula**. A radula is structure with many teeth attached to an **odontophore** that is found in most of the mollusc groups.

Class **Scaphopoda** are the tusk shells. These organisms have a single cone-shaped shell with an opening at each end to allow water to pass over the mantle for gas exchange. They burrow into soft sediment and capture prey e.g., foraminifera, with their tentacles. A scaphopod radula has flattened teeth modified for grinding food into smaller pieces to make it easier to digest. Tusk shells are highly valued and traded as a type of currency by many western Native American tribes. The shells are strung together to make necklaces and other decorative items.

Class **Gastropoda** includes the snails, limpets, abalone, slugs and nudibranchs. The snails have a single shell that coils and increases in size as the animal grows. These shells may include all kinds of adaptive features such as rows of spines that help the organism protect itself from predators. Some gastropods like limpets and abalone have single shells that do not coil. These shells are often low and flattened to allow the animal to be protected from predators but also resist other kinds of forces such as heavy wave action. Terrestrial slugs and nudibranchs (sea slugs) have lost their shells and rely on other adaptations for protection. Some gastropods are herbivores and use their radula for grazing much like the chitons. Others are carnivores, scavengers or filter-feeders and adaptations for this lifestyle are evident in the different forms that their radulas take. Some have teeth that are longer and sharper for ripping and tearing the flesh of their prey others have modified their radulas into drill-like structures that can work their way through the shell of a potential meal.

Class **Bivalvia** are the clams, oysters and mussels. They are characterized by having a symmetrical two-part shell that is hinged. Bivalves are the only molluscs that do not have a radula. Most bivalves are filter feeders. They bring water into the mantle cavity using an incurrent siphon. The water passes over their gills and as gas exchange occurs, small organisms are also trapped. The trapped food items are transported along ciliated-tracks to the mouth of the organism. The remaining water is then passed out of the mantle cavity through the excurrent siphon. Many experts are able to dig for the particular species of bivalve they like to eat by looking at the 'show'; the distinctive look of the paired siphons poking up out of the sand. Not all bivalves burrow into the sand, some sit on top of the sediment surface, e.g., scallops, some attach to surfaces using adaptations like tough proteinaceous byssal threads e.g., mussles, some bore into soft sedimentary rocks e.g., piddock clams, and some can even bore into wood. In the days of wooden sailing ships the ability of these organisms to burrow into wood and digest the cellulose with the aid of bacterial symbionts often determined the life expectancy of a ship.

Class **Cephalopoda** are the nautiloids, cuttlefish, octopods and squids. The nautiloids are the shelled cephalopods. They are represented by the ammonites, an ancient group that is very prevalent in the fossil records, so much so that they are used as reference species to try to verify the age of sediment layers. Several species of *Nautilus* can still be found in the deep waters of the Indo-Pacific Ocean. The *Nautilus* has a complex coiled shell with many chambers. It has the ability to control its buoyancy much like a submarine, by using osmosis to control which of the chambers contains gas and which contains water. These organisms are active swimming predators that swim using jet propulsion and catch their prey with their tentacles. Cuttlefish have lost their outer shell but still have a very porous 'cuttle bone' which is really a shell embedded in their mantle that assists the organism with buoyancy. Cuttle bones are often found in the bird section of pet stores, and are given to some species of caged birds to grind their beaks on. Squid have reduced their hardened parts even further until all that are left are a thin, flexible proteinaceous pen embedded in the dorsal side of the mantle, a hardened beak-like structure around the mouth, and in some species toothed-rings around the edges of the suckers of the tentacles. The accompanying reduction in the weight supports a very active predatory pelagic lifestyle. Octopods have only a hardened beak and many, although predatory, have returned to the benthos. Cephalopods are carnivores and this is reflected in the shape and size of the teeth found in their radula. Many species of cephalopod are able to change their appearance dramatically for protection and communication by being able to manipulate small areas of pigment on their skin called **chromatophores,** which allows them to rapidly change the color of all or part of their body. A sac filled with an inky substance that can be discharged into the water to confuse a predator is another interesting feature found in cephalopods. Cephalopods have excellent vision and are extremely intelligent. In this lab you will become familiar with the anatomy of a representative bivalve and cephalopod by carrying out a careful dissection.

Materials for Each Group

- goggles
- gloves
- aprons
- dissecting tools (scissors, probes, scalpels, forceps)
- dissecting trays
- Squid dissection
- representative bivalve for dissection
- tissue disposal bucket
- clean-up materials
- compound microscope
- prepared radula slide
- clean slides and cover slips

Exercise 1
Bivalve Dissection

In this exercise you will investigate the anatomy of a representative bivalve.

1. Get into groups of 3 to 4 students. Collect your specimen, dissection tools and safety equipment. Always wear eye protection when dissecting.

2. Examine the outer part of the specimen. Note the symmetry of the shells around the hinge. The area of the hinge is the dorsal surface of the animal. Refer to Figure 8-1 and locate the 'bump' in the dorsal surface of the shell that is the **umbo**. Use the umbo as a reference point to locate the anterior (head) and the posterior (back) portion of the organism.

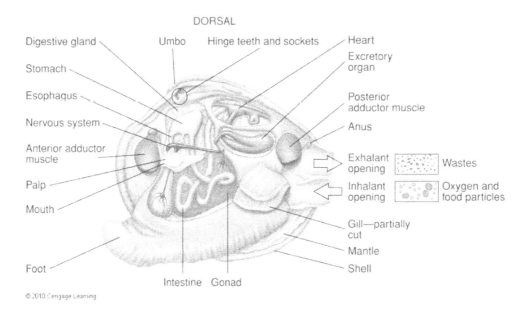

Figure 8-1 Bivalve Anatomy.

3. Using your dissection tools carefully open the shell and expose the animal. Use Figure 8-1 as a reference to identify structures. The mantle will be adhering to the shell. An **adductor muscle** at the anterior and another one at the posterior end of the animal work to keep the shell closed when the animal is alive. A pair of **gills** hang will be hanging ventrally in the **mantle cavity**. Use a blunt probe to lift up the gills and expose the rest of the body (**visceral mass**).

4. Locate the **incurrent (inhalant)** and **excurrent (exhalent) siphons** at the posterior end of the organism. See if you can locate the mouth and the foot at the anterior portion of the organism

5. Bivalves do not have a complete closed circulatory system like humans but they do have a two-chambered **heart** that lies on the dorsal surface of the visceral mass in a pericardial sac. Use a blunt probe and try and locate the heart.

6. Remove the mantle and the gills and make an anterior to posterior cut in the visceral mass. Try and locate the stomach. If your specimen is large and fresh enough you may find a hard, rod-like structure inside the stomach. This is the crystalline style that aids in digestion. Specimens that have spent a long time without feeding will absorb the structure.

7. Draw and label your specimen in the space provided. **When your observations are complete make sure all material is disposed of in the tissue disposal bucket provided.**

Exercise 2
Dissection of a Representative Cephalopod

In this exercise you will investigate the anatomy of a representative Cephalopod, the squid.

1. Get into groups of 3 to 4 students. Collect your specimen, dissection tools and safety equipment. Always wear eye protection when dissecting.

2. Examine the exterior of the specimen. Using Figure 8-2 as a reference, locate the anterior and posterior portion of the organism. Note the stream-lined mantle with the fins at the posterior end. Lift up the mantle and note the mantle cavity. Lay the animal on its dorsal side with the ventral surface facing up and locate the **funnel**. The animal moves by jet propulsion because it is able to fill its mantle cavity with water, contract the muscles and the water will shoot out the funnel.

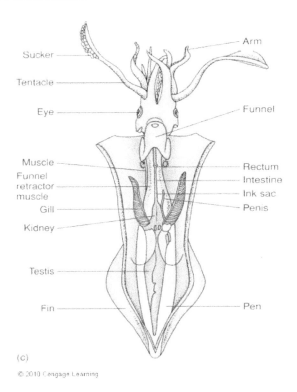

(c)

© 2010 Cengage Learning

Figure 8-2 Squid Anatomy

3. Turn the squid over and examine the dorsal surface. At the midline of the mantle you will be able to feel a stiff structure. You may be able to pull on the end of the structure and remove it-if you can not-use a scalpel and make a shallow incision to expose the structure. Once you have exposed it pull it out-this is the **pen**. The pen helps stabilize the body.

4. In the area of the mantle where you have already made a shallow incision to remove the pen, peel off a small section of the mantle skin and place on a microscope slide to observe the areas of pigment called **chromatophores**. Note whether the dorsal or the ventral surface of the organism is more heavily pigmented.

5. Note the **8 arms** and 2 **tentacles** that circle the mouth region of the head. Locate the **suckers** on the surface of the arms and tentacles and see if there is an extra ring with hooks and remove one of them for further observation. These hooks help the squid to capture its prey.

6. Pull back the arms of your specimen and locate the ball of muscle around the mouth area that is the **buccal mass**. Inside the buccal mass will be a hardened **beak**-like structure that can be removed. Inside the beak you will be able to locate the **radula**.

7. Remove the radula and place on a microscope slide and examine further using a compound microscope. Compare your specimen to the prepared slide of a radula.

8. Return to your whole specimen. Lay it on its dorsal surface. Using a pair of scissors cut the mantle cavity open down the midline of the ventral surface. Be careful not to go too deep or you will damage the internal organs. Lay open the cut mantle and observe the organs.

9. Locate the paired **gills** in the body cavity. The squid is a fast-swimming predator and needs a good oxygen supply to support its active lifestyle. Unlike the bivalve which has an open circulatory system, the squid has a closed circulatory system and three hearts (two branchial hearts, one at the base of each gill and one systemic heart).

10. Locate the **ink sac** on the surface of the visceral mass.

11. Squid reproduce sexually and have separate sexes. You may be able to distinguish the sex of your specimen. If you have a mature female you may be able to see the **nidamental glands**, which are paired white structures that produce the covering for the eggs, ovaries or oviducts containing eggs. If you have a mature male you may be able to locate the **testes** at the posterior end of the mantle cavity and the spermatophoric sac. Males produce sperm in packets that they transfer to the mantle cavity of the female using a special structure on the fourth arm called a **hectocotylus**.

12. Draw and label your specimen in the space provided. **When your observations are complete make sure all material is disposed of in the tissue disposal bucket provided.** Draw and label your squid dissection in the space provided.

Clean and dry all dissection tools carefully. Wipe down work space. Wash your hands thoroughly before you leave the lab.

Questions

1. Do all molluscs have a radula?

2. Do all molluscs have a shell?

3. Are there any molluscs that can damage wood?

4. If your squid specimen was more darkly pigmented on one side compared to the other, what purpose would this serve?

5. What is the advantage to shell reduction in the cephalopods?

Suggested Readings

Museum of Victoria, 2008. Public dissection of a giant squid.
http://museumvictoria.com.au/about/mv-news/2008/giant-squid-public-dissection-at-melbourne-museum/

Ruppert, E. E., Fox, R. S. and R. D. Barnes. 2004. Invertebrate Zoology: A Functional Evolutionary Approach, 7th Edition. Brooks/Cole Publishing Co. 1008 p.

Glossary of Terms

anterior: Anatomical term for the front end of an organism, opposite to posterior.

bilaterial symmetry: When an organism has two matching sides.

byssal gland: Gland inside a mussel that secretes tough attachment threads called byssus.

dorsal: An anatomical term for the upper side of an organism.

chromatophores: Special pigment containing cells found in the skin of some organisms.

crystalline style: A rod-like structure that both aids in mechanical and enzymatic digestion in some bivalves.

mantle: Part of the body wall that covers the visceral mass in molluscs; secretes the shell.

pen: A thin flexible strip of proteinaceous material that is embedded in the mantle of a squid.

odontophore: Tongue-like muscular structure on which the radula is attached.

posterior: An anatomical term for the back end of an organism, opposite to anterior.

radula: Toothed structure for feeding, unique to molluscs.

siphon: Tubular structures found in some organisms e.g., clams, that moves water in and out of the body cavity.

umbo: A prominent area of shell found at the hinge area in a bivalve.

ventral: An anatomical term for the lower or under side of an organism.

Lab 9

Macroalgae

Objectives

Upon completion of this lab exercise, you should be able to:

1) Describe the function of **primary producers** in the marine ecosystem.

2) Identify the basic macroalgal structures and describe basic differences between terrestrial plants and algae.

3) Identify different **pigments** characteristic of different groups of algae.

4) Identify the **abiotic factors** that determine the location of primary producers.

5) Discuss how algae are used in our daily lives.

Background-Macroalgae

Most forms of life depend either directly or indirectly on the organic materials produced by primary producers. Primary producers utilize carbon dioxide and water to produce organic molecules from inorganic materials during the photosynthetic process. Scientists measure primary productivity in terms of carbon "fixed" or converted into organic material. The unit of measurement of primary productivity equals the number of grams of carbon per square meter of surface area per year (g $C/m^2/yr$). The oceans' primary productivity averages from 75 to 150 g $C/m^2/yr$ (35–50 billion metric tons/year), while the land's primary production rate is about 50–70 billion metric tons of carbon per year. Most of the primary production in marine ecosystems is carried out by the phytoplankton (photosynthetic bacteria and microalgae) but macroalgae (seaweeds) and marine plants are also important.

Macroalgae and marine plants are important to marine ecosystems because of the oxygen they produce and the carbon they fix during photosynthesis. They also provide important structural complexity. They provide food, places for lots of organisms to hide and attach, and protection from wave action. For example, the Kelp ecosystems, characterized by large stands of brown algae, can harbor over 1000 different species, many of them important to commercial fisheries. Increasingly marine habitat restoration efforts are focused on the health of the algal and estuarine plant communities.

The distribution of macroalgae is affected by the amount of light transmitted through the water column, temperature, nutrient availability, pressure from herbivory as well as factors such as the duration of tidal exposure, strength of wave action and type of substrate available for attachment. There is a great deal of variability in seaweed structure and this can reflect adaptations for living in different habitats. Unlike terrestrial plants, seaweeds do not have the same kind of vascular structures for transporting food and water-although some species of kelp (e.g., *Macrocystis*) do have some modified cells for transport (not surprising since they can be over 60 m long). So while seaweed structures may bear a passing resemblance to the roots, stems and leaves of plants they are different. The **thallus** (plural thalli) is the seaweed body. The flattened portion of the thallus is called the **blade**. The blade may be suspended on a stem-like structure called the **stipe** and attached to the substrate by **holdfasts**. Some species may also have **airbladders** (also called pneumatocysts) for suspending the stipe in the water column (Figure 9-1). Some species also have reproductive structures that are visible, for example Rock Weed (*Fucus* sp.) produces bulbous reproductive structures at the tips of branching thalli. Sea Palms (*Postelsia* sp.) are found on rocky outcroppings that receive heavy surf action. They are able to live in this environment because of a very flexible stipe and strong holdfast. Other seaweeds have body-types that are low or have cell walls containing calcium carbonate ($CaCO_3$), which also allows them to withstand significant force and resist herbivory.

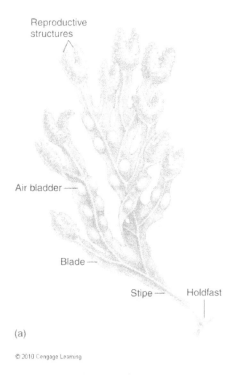

Reproductive structures

Air bladder —

Blade —

Stipe — Holdfast

(a)

© 2010 Cengage Learning

Figure 9-1 Seaweed structure e.g., *Fucus*

The portion of the electromagnetic spectrum referred to as **visible light** (380-750 nm) is made up of light of different colors (wavelengths) Figure 9-2.

UV	Violet		Blue	Green	Yellow	Orange	Red	infrared
400	450		500	550	600 650		700	750nm

Figure 9-2 The wavelengths associated with visible light

Photosynthetic organisms are able to capture this energy by absorbing light of particular wavelengths using photosynthetic pigments. Algae and terrestrial plants contain chlorophyll as well as other characteristic accessory pigments. **Chlorophyll a** is the major photosynthetic pigment and is found in all photosynthetic algae and plants, but there are other forms of Chlorophyll (b, c and d). **Chlorophyll a** absorbs light most effectively at about 430 nm (violet) and then again at 662 nm (red light). **Chlorophyta** (green algae) also have **chlorophyll b** (peak absorbance 453 nm ,blue and 642 nm, far red) and a group of accessory pigments called carotenoids (carotenes and xanthophylls), many of which they share with the closely related terrestrial plants. Rhodophyta (red algae) also have **chlorophyll d** and the **phycobillins** (**phycoerthythrin and phycocyanin**) which can absorb more light in the green range. **Phaeophyta** (brown algae) are characterized by **chlorophyll c** and a type of carotenoid pigment called **fucoxanthin** (absorbs in the blue-green and yellow green light range). All of the different pigments absorb light of slightly different wavelengths and some of them are photoprotective (i.e., sunscreen). As light penetrates ocean waters some of the wavelengths are absorbed sooner than others. In clear water, red and yellow wavelengths of light are absorbed quickly but blue light penetrates to deeper depths. However in water that is more turbid, i.e., with more suspended particles, sometimes the green wavelengths of light penetrate more deeply, as in coastal waters. It was once thought that the type of accessory pigments determined the depth at which a particular type of algae would be found, there is a correlation but other factors such as herbivory and competition are thought to be even more significant in determining where a particular species will be found.

Seaweeds or macroalgae have been used by humans for centuries. They are an important source of food in many countries and commercial seaweed operations are economically significant particularly in countries like Japan. Some species are a good source of iodine or vitamin C and were used by coastal people to prevent goiter and scurvy. The nori that wraps that outside of sushi is a type of red algae called *Porphyra,* also known as Laver and found in Welsh laver bread. Other commonly eaten forms of seaweed include dulse, kombu and wakame. The cell walls of red and brown algae contain **phycolloids** that help protect the organism from drying out but are economically valuable for their gelling properties, e.g., **agar** and **carrageenen**. Agar is used extensively to make microbiological media and further purified as agarose, used for gel electrophoresis. A grocery store search for carrageenen would find it in the ingredient list of many products including ice cream. Scientists are actively researching the potential for algae to be important in bioremediation and as a source of biofuel.

Materials

- assorted examples of green, red, and brown algae; fresh specimens, herbarium specimens or photographs can be used
- hand lens
- compound microscope
- mortar and pestle
- goggles and gloves
- approx 10-20 mL of 20% ethanol/80% acetone solution (take from stock
- bottle and put in a small beaker
- marine algae sample (see instructor for how much) (Chlorophyta and/or Phaeophyta)
- fresh spinach leaves (or other dark green representative terrestrial plant leaves)
- unknown algal extract (prepared ahead of time)
- funnel
- filter paper (coffee filters work just fine)
- beaker of appropriate size to balance funnel
- capillary tubes
- chromatography paper
- Erlenmeyer flask
- Large test tube with cork cap with paper clip hook, set up with chromatography solvent
- chromatography solvent (NOTE this contains ether-do NOT leave uncapped at any time)
- pencil
- colored pencils
- grease pencil

Optional Field Exercise Materials:

- $0.5m^2$ quadrat

- 100 m horizontal transect rope

- 50 meter vertical transect rope

- clipboards

- pencils

- GPS (optional)

- reference charts and identification keys

Safety Notes

In this lab, if you are working with preserved samples of algae and your samples have been shipped in a preservative that contains ethylene glycol and isopropanol, it is strongly recommended that you wear glasses instead of your contact lenses. These chemicals have been known to cause eye irritation. You should also wear gloves when handling the preserved specimens. Maintain good ventilation in laboratory area when working with organic solvents for pigment extraction. Chromatography solution contains petroleum ether, do not leave filled tubes uncapped in laboratory. Adding chromatography solution to tubes or emptying them must take place in a fume hood. Petroleum ether and acetone are flammable, avoid sparks.

Exercise 1
Diversity in Macroalgae Structure

1. Using the specimen materials available in lab (e.g., herbarium, preserved or fresh specimens) and the space provided below draw and label at least one example of Chlorophyta, Rhodophyta and Phaeophyta.

   Make notes if structures are adaptations that indicate a particular habitat (e.g., long stipe may indicate specimen was found in deeper water)

Example Chlorophyta Genus if known: Common Name if known: Collection location:	Example Rhodophyta Genus if known: Common Name if known: Collection location:	Example Phaeophyta Genus if known: Common Name if known: Collection location:

Exercise 2
Indentifying an Unknown Algal Sample using Paper Chromatography

The different groups of algae are, in part, classified based on their photosynthetic pigments. All photosynthetic algae contain chlorophyll a. Chlorophyta share many of the same pigments as terrestrial plants but Phaeophyta contain fucoxanthin and Rhodophyta contain phycobillins. In this lab exercise you will use a technique called paper chromatography to separate the pigments found in plant and algal samples. When you put a sample of an extracted pigment on the special chromatography paper and let a solvent wick up the paper, the different pigments will separate and travel different distances. The distance a pigment travels will depend on its size and how soluble it is in the solvent. Small non-polar pigments will travel the fastest while large polar pigments will tend to stay close to the point of origin. The characteristic pigments found in Rhodophyta are water soluble and are not suitable for examination with this chromatography technique. You will extract the pigments from two different types of algae and a land plant and use the paper chromatography results as a reference to identify an unknown algal extract.

Read directions ALL the way through before beginning.

1. Get into groups of 3-4 students. Gather your materials, make sure you are wearing appropriate safety gear.

2. Place a sample of green algae (Chlorophyta) into mortar and pestle

3. Add enough 20% ethanol/80% acetone to cover the tissue

4. Grind until you have created a very smooth paste and the solvent has changed color.

5. Use funnels, and filters provided with an appropriately sized beaker to set up a filtering station. Make sure you label collection container.

6. Scrape the contents of your mortar and pestle into the filter. You can rinse any adhering pieces with some of the 20% ethanol/80% acetone solvent. Collect the liquid that passes through the filter. This is now your pigment extract.

7. Discard filter and clean funnel when completed-so it can be used again for the other samples.

8. Take a strip of chromatography paper and trim one end to a point as shown below. Try to only touch chromatography paper on the edges because the oils from your hands will interfere with the procedure.

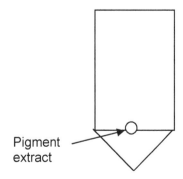

Pigment extract

9. Use a capillary tube to transfer a drop of the pigment extract to the chromatography paper approximately 1.5 cm above the point. To make the pigment spot as dense as possible you will need to reapply 30-40 times-letting it dry in between. **Be patient!**

10. **Repeat** extraction procedure steps 2-10, with a sample of brown algae (division Phaeophyta), a plant sample e.g., spinach, and the unknown, so spots can be applied at the same time (to different strips of chromatography paper) for efficiency

11. When you have made a dense pigment dot, attach to the chromatography tube apparatus. Dispose of capillary tube in appropriate container DO NOT throw in the trash.

Only the tip of the chromatography paper should touch the solvent.

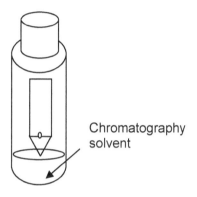

Chromatography solvent

14. Cap chromatography tube and set in Erlenmeyer flask or test tube rack for balance.

15. Allow to stand for approximately 5 minutes and then remove chromatography paper from the chromatography apparatus and observe. Make sure the solvent front does not reach the top of the chromatography paper or you may lose the beta-carotene line.

16. Using the characteristics given in Table 9-1 try and identify the different pigments.

Table 9-1 Pigment characteristics

Pigment type	color	molecular wt g/mol	expected in
Beta-carotene	chrome yellow	536.88	brown and green algae, land plants
Xanthophylls (lutein)	yellow	568.88	brown and green algae, land plants
Xanthophylls (fucoxanthin)	orange	658.91	brown algae
Chlorophyll a	blue-green	893.5	brown and green algae, land plants
Chlorophyll b	yellow-green	907.47	green algae, land plants
Chlorophyll c	light green	598 (variable)	brown algae

Using colored pencils draw and label your Chromatography results below. Try to identify the different pigments.
Based on your results what kind of algae was your unknown? _____

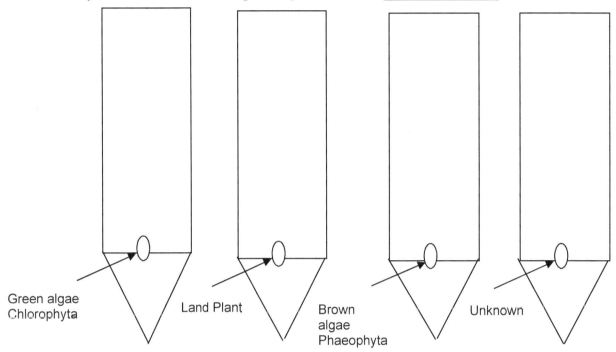

Green algae
Chlorophyta

Land Plant

Brown
algae
Phaeophyta

Unknown

Exercise 3
Macroalgae Identification

This is a field option. Your instructor may customize the directions to suit the resources available to your class.

1. Randomly choose three sample areas at each of these tidal zones: 1-10 m, 15-20 m, and 25–30 m from the tidal zone. Identify how many of the following algae are present in each of the zones by placing the number of organisms on Table 9-2. (You may use an actual study site at the beach, near a fresh water riverbank, or pictures taken from an ocean tidal zone) You can use sampling quadrats and random sampling techniques in your selected sample area.

2. Place a transect line along your study area and then randomly select areas that are along a vertical transect within your study area.

3. Your instructor will provide identification keys appropriate to your area and site

4. Record the number and type of each algal species found in each of quadrat or picture samples.

Group Names:

Site Sampled:

Coordinates:

Date Collected:

110 Macroalgae

Depth from High-Tide Line	Sample 1 a. (1-10m)	b. (15-20m)	c. (25-30m)	Sample 2 a. (1-10m)	b. (15-20m)	c. (25-30m)	Sample 3 a. (1-10m)	b. (15-20m)	c. (25-30m)
Algae Species									
1.									
2.									
3.									
4.									
5.									
6.									
7.									
8.									
9.									
10.									
11.									

Exercise 4
Macroalgae Monitoring

California Rocky Coast Algae Identification

Use the Limpets Monitoring Rocky Intertidal website (http://limpetsmonitoring.org/ri_algae.php)

to answer the following questions about the different algae found on the California coastline.

Click on species monitored. Your instructor may ask you to use local resources or update the link.

Name of Algae	Description of Structure Used To Identify It: Brown, Red, or Green?	Distribution and Habitat	Why Is It Monitored?	Fun Facts About the Algae
Feather Boa Kelp *Egregia menziesii*				
Green Pin-Cushion *Cladorphora columbiana*				
Dead Man's Fingers *Codium fragile*				
Sea Lettuce *Ulva*				
Surf Grasses *Phyllospadiz scouleri*				
Flattened Rockweeds *Fucus gardneri*				
Slender Rockweeds *Pelvetiopsis limitata*				
Tar Spot Algae *Mastocarpus* spp.				

Name of Algae	Description of Structure Used to Identify It: Brown, Red, or Green?	Distribution and Habitat	Why Is It Monitored?	Fun Facts About the Algae
Encrusting Coralline Algae				
Upright Coralline Algae *Bossiella* spp.				
Scouring Pad Algae *Endocladia mricata*				
Stunted Turkish Towel *Mastocarpus* spp.				
Nori *Porphyra spp.*				
Sea Sacks *Halosaccion glandiforme*				
Lawn Algae *Chondracanthus canaliculatus*				
Iridescent Algae *Mazzaella flaccida*				

Questions

1. How are the basic algal structures (e.g., blade, stipe, holdfast) different from the leaves, stems and roots of land plants? What are the functions of those structures?

2. How were you able to identify the type of algae that was the source of the unknown pigment extract?

3. Why do green algae and terrestrial plants share so many of the same photosynthetic pigments?

4. What is the purpose of accessory pigments?

5. How are the different pigments found in different types of algae related to the depth at which each algae is found?

6. Why is it necessary for biologists to monitor the algae growth on the coastline?

7. Other than food sources for ocean communities, what are other ways these different forms of algae are used in our society?

8. Use your text and other resources to fill in the table below as a review exercise

Type of Algae	Main Pigments	Accessory Pigments	Chemicals Making Up Cell Walls	Evolutionary Ancestors	Ocean Zone in which They Are Found
Green					
Red					
Brown					

Glossary of Terms

abiotic: Non living ecosystem factors e.g., temperature, salinity.

air bladder: An air sac found on algae that contains carbon dioxide gas to allow the plant to remain upright.

blade: Flattened leaf-like portion of algal thallus.

chlorophyll: Photosynthetic pigment, small functional groups substitutions can alter absorption spectra for the different forms, chlorophyll a, b, c and d.

desiccation: The drying out of organisms due to exposure to sunlight and high levels of salt.

holdfast: An algae appendage that anchors the organism to rocks or other substrate.

phytoplankton: Single-celled photosynthetic organisms that are free-floating.

phycolloids: Polymers of sugars found in the cell walls of different types of algae e.g., alginates, carrageenan and agar.

pneumatocyst: An air sac found in brown algae containing carbon dioxide gas to keep the algae upright in the tidal region.

primary producer: An organism, found at the beginning of a food chain, that utilizes sunlight or chemosynthesis to make organic compounds.

quadrat: A sampling tool used to mark a specific area of study. It can be made out of PVC piping and string to mark a square meter in the natural environment. For pictures, you could use popsicle sticks to mark an area one-tenth the distance for easier sampling in the classroom.

stipe: The stem-like structure of the algae to which the blades attach.

substrate: The material that makes up the bottom of the ocean floor. It can be sand, rock, coral, or other manmade materials.

thallus: The complete structure of the brown algae made up of blades, stipes, and holdfast. Unlike terrestrial plants, the complete structure is photosynthetic.

Lab 10

Fish Niche

Objectives

Upon completion of this lab exercise, you should be able to:

1) Propose questions after you make initial observations of the fish in single species and community tanks.

2) Generate tentative hypotheses about the fish communities.

3) Design your own experiment, identifying an **independent** variable and a **dependent** variable.

4) As part of your experimental design, your group will formalize a hypothesis with predictions.

5) Communicate your results in a format specified by your instructor.

Communities and Niches

An ecological **community** is an interacting group of species. Some communities have high **species diversity**, which means that they are composed of many different species. Terrestrial and aquatic communities in the tropics often are wonderfully diverse, with many interacting species. Other communities, such as coniferous forests in the temperate zone, naturally have lower species diversity. Unfortunately, many types of communities are rapidly losing their original species diversity. While the causes of this loss of species are as complex as the communities themselves, they are sometimes summarized with the acronym HIPPO, which stands for a set of anthropogenic (human-caused) factors:

1) Habitat destruction

2) Introduced species

3) Pollution

4) Population growth

5) Over exploitation

In order to improve, or even reverse, the effects of these destructive factors we need to understand the inner dynamics of communities. What holds ecological communities together? What breaks them down? What allows species within a community to coexist?

Coexistence is the ability of multiple species to live in the same community without one species driving another to local extinction. One phenomenon that allows coexistence may be that each species within a community uses a characteristic and unique set of resources for food and shelter.

A **niche** represents the ways a species interacts with its environment to obtain all the resources it may use for food, shelter, reproduction, and any other activity that is characteristic of the species. A niche is sometimes described as the role that the species plays in the community. Complex communities with high species diversity often contain sets of closely related species that occupy different and highly specialized niches. The cichlid fishes in the great East African rift lakes of Victoria, Tanganyika, and Malawi are a striking example of extraordinary species diversity coupled with extremely specialized niches. Stiassny and Meyer (1999) reported that each of these lakes contains from 200 to 500 different cichlid species, although the species richness of Lake Victoria is now reduced due to the introduction of invasive species.

Competition between members of different species, **interspecific competition**, is harmful to both species because it makes it more difficult for them to access resources such as food or shelter. The development of specialized and narrower niches that do not overlap reduces interspecific competition, and by doing so, allows them to coexist. This process is called **niche partitioning**. It may play a fundamental role in promoting the coexistence of similar species, and in allowing for the development of complex communities with rich species diversity.

Niche partitioning can result in permanently increased morphological and behavioral differences between species. Alternatively, niche partitioning can result in more flexible adjustments that individuals of one species make only when they are in the presence of individuals of a second species that is a potential competitor. McCauley (et al. 2012) recently documented a shift in the habits of fish on a coral reef where over fishing had reduced the predator population. Fish species that were active mostly at night in reefs with abundant predators were also active during the day in reefs where the predators were absent.

When niche partitioning is a result of adjustments made by individuals to avoid competition for resources in the presence of a second species, each species should use a more specialized set of resources in communities that contain similar species than in communities that do not contain these potential competitors. The niche of a species in the absence of potential competitors is called its **fundamental niche**; the niche of a species in the presence of potential competitors is called its **realized niche.**

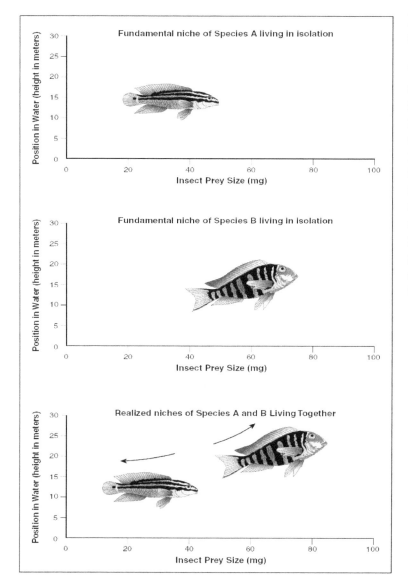

Figure 10-1 Fundamental and Realized Niches.

Two fish species that eat aquatic insects shown in relation to prey size and foraging location in the water column. The drawings of the fish represent the height in the water column at which they are foraging, and the size range of prey they are capturing. The arrows in the third panel represent niche partitioning, which may occur when the species are in the same community.

Materials

For the whole class:

- Four fish tanks (three containing one species of fish only and one tank containing all three species of fish)

- Three types of fish, all either freshwater or saltwater

- Access for each group to white board or black board space or chart paper for posters

- Stopwatch (optional)

- Small clear metric ruler (optional)

- Additional materials may be requested from your instructor to modify the tank environment in non-destructive ways

Live Organisms: Care and Handling Review (summary from lab 1)

Live organisms are commonly found in laboratory learning environments.

- All live organisms used in laboratories should receive the highest standard of care.

- Organisms may have an entirely different level of sensory perception and development than humans, and appropriate care must be taken when handling and caring for these organisms not to disturb them.

- Many organisms commonly found in educational laboratories may not be native, do not release them to the environment.

Exercise 1
Fish Behavior

Work with your group to propose and evaluate a hypothesis about the community ecology of fishes by working through the following stages of this exercise

1. Get into groups of 3-4 students. Propose questions as you make initial observations of the fish in single species and in community tanks. After producing a set of initial observations and questions, your group will brainstorm to generate tentative answers (hypotheses) about the fish communities.

2. After your group has formalized a testable hypothesis, you will design your experiment. The experimental or **independent** variable of your experiment will be the number of fish species in the tank: one species, or multiple species. The **dependent** variable will vary between groups as each group will be testing a different hypothesis. As part of your experimental design, your group will generate predictions.

3. Observe the fish in different tanks as you collect your data.

4. Present, with your group, a brief summary of your experiment to the class.

5. Individually, write a lab report as directed by your lab instructor.

Hypothesis Formation

This is the most creative and challenging stage of your experiment. Your group will work together to come up with an interesting question and tentative answer, a hypothesis, about the community ecology of fish. In order to be testable within the constraints of this exercise, it will need to lead to predictions about fish in the single-species and community tanks present in the laboratory. First, you will spend about five minutes at each tank recording your initial observations of, and questions about, fish in the single-species and community tanks. Observing animals in a group can be challenging. Consider focusing on one animal in the group. Then repeat your observations using additional members of the group so you can make generalizations about the behavior. If you cannot distinguish one individual from another, you might choose an arbitrary time interval (e.g., every 15 seconds) and try to record what all the fish are doing at that instant. Do not tap the tanks or otherwise disturb the fish during your initial observations. If you are having trouble generating interesting questions, consider the general prompts listed below. Secondly, your group will discuss your initial observations and questions and use them to generate hypotheses.

- Do you see evidence of partitioning of resources (i.e., different niches)?
- How are the fish relating to members of their own species?
- How are they relating to members of other species in the community tank?
- How are they distributed within the tank?
- If they are staying still, where are they positioned?
- If they are moving, how do they move?
- What stimulates them to move?

Table 11-1 Fish Species

Species	Common Name

Species 1

Initial Observations and Questions

Species 2

Initial Observations and Questions

Species 3

Initial Observations and Questions

Community Tank

Initial Observations and Questions

Review the observations and questions of other members of your group, and as you do so, record hypotheses in the space below. Once you have a list of hypotheses, your group needs to decide which one is most interesting and testable in this laboratory. Underline the hypothesis that your group has decided to test.

Hypotheses:

Experimental Design

To test your hypothesis, you will determine if the independent variable influences the dependent variable in the manner predicted by the hypothesis. The independent variable, the variable that has been manipulated, is the number of species in a tank. It has been set at two levels: one species, or multiple species. What is your group's dependent variable, and how does your hypothesis predict it will respond to variation in the number of fish species in the tanks?

Description of Your Dependent Variable:

Predicted Result of Your Hypothesis:

Observed Results

You are ready to collect your observed results so that you can compare them to the predictions of your hypothesis. Notice that it would not be a fair test of your hypothesis to use your initial observations as observed results, as almost any hypothesis could be constructed to accurately predict what has already happened. Record your observed results of fish in single- and multiple-species tanks on the following pages. Remember that observing animal behavior requires patience and attention to detail.

Observed Results of Fish in Single-Species Tank:

Species 1:

Observed Results of Fish in Single-Species Tank:

Species 2:

Observed Results of Fish in Single-Species Tank:

Species 3:

Observed Results of Fish in Community Tank:

Species 1

Species 2

Species 3

Conclusions

Use a white board, chart paper or the blackboard to prepare a poster that will illustrate your hypothesis, how it leads to the predicted results, and how they have compared with your observed results. Circulate around the room and observe each group's poster. Record their information so that you can cite their work as you write your discussion.

Questions

Write a lab report including the following information:

1. Write an introduction. In this introduction, present your hypothesis, explain its significance, and briefly give an overview of how you tested it.

2. Write a methods section. Be sure to clearly explain the details of your methods. For example, how much time did you spend at each tank?

3. Write a results section. This section may, or may not, contain a graph or table, but it must contain text describing the response of your dependent variable to the independent variable.

4. Write a discussion including both a conclusion and implications. In your conclusion, describe the relationship between your hypothesis and its predicted result and compare this prediction to your observed results.

5. Explain what this comparison suggests about your hypothesis.

6. In your implications, consider what your conclusion tells us about the structure of communities in general. How does your conclusion compare to the results and conclusions of other laboratory groups?

7. How could people apply an understanding of the structure of natural communities to a specific type of issue in conservation biology?

Suggested Reading

McCauley, D. J. Hoffmann E. Young, H. S. and F. Micheli. 2012. Night shift: expansion of temporal niche use following reductions in predator density. PLoS ONE 7(6): e38871. doi:10.1371/journal.pone.0038871

Stiassny M.L.J. and A. Meyer. 1999. Cichlids of the Rift Lakes. *Scientific American*, 280: 64–69.

Stiling, P.D. 1992. Ecology: Theories and Applications. *Prentice Hall*, Upper Saddle River, NJ.

Weiner, J. 1995. The Beak of the Finch. *Random House*, New York.

Wooton, R.J. 1990. Ecology of Teleost Fishes. *Chapman and Hall*, London.

Glossary of Terms

community: An assembly of populations of different species that occupy the same habitat at the same time.

fundamental niche: The broadest of all possible niches that an organism theoretically can occupy.

specialist species: Species with a narrow ecological niche. They may be able to live in only one type of habitat, tolerate only a narrow range of climatic and other environmental conditions, or use only one type or a few types of food.

interspecific competition: Competition between members of a single species for a limited resource.

niche: An organism's role in its environment.

niche partitioning: The changes in niche breadth by species to inhabit specialized niches, so as to reduce interspecific competition.

realized niche: The niche that an organism actually occupies. It is narrower than an organism's fundamental niche because of interspecific competition.

species: One or more populations of potentially interbreeding organisms that are reproductively isolated from other such groups.

species diversity: Number of different species (species richness) combined with the relative abundance of individuals within each of those species (species evenness) in a given area.

specialist species: Species with a narrow ecological niche. They may be able to live in only one type of habitat, tolerate only a narrow range of climatic and other environmental conditions, or use only one type or a few types of food.

Lab 11

Animal Behavior

Objectives

Upon completion of this exercise, you should be able to:

1) Observe and be able to recognize Hermit crab shell acquisition behavior.

2) Discuss the role of **visual**, **mechano**-and **chemoreceptors** in shell acquisition behavior.

3) Using Hermit crabs or an animal approved by your instructor, design an experiment to investigate the primary sensory receptor used by a marine organism to respond to cues in its environment.

4) Identify the independent and dependent variables in the experiment.

5) Formulate a hypothesis about the relationship between experimental variables, and make a prediction regarding the experimental outcome.

6) Conduct the experiment and collect data. Summarize the data in a table and graph the results.

7) Use the results to evaluate the hypothesis and draw conclusions.

8) Summarize your results in writing and submit to your instructor as instructed.

Hermit Crab Behavior

Animals in all environments share some common challenges. They have to find food, find a mate, find shelter and avoid predators. Complex behaviors have evolved to let an animal respond appropriately to the information received by its sensory structures. One of the dominant animal groups on the planet is the arthropods with at least three-quarters of a million described species. Arthropods have undergone a diversification associated with niche specialization that lets them take advantage of many different habitats. Arthropods are characterized by an exoskeleton and segmented appendages. The sensory receptors, that let arthropods gather and respond to information in their environment, are modifications of the exoskeleton. In this lab you will investigate the shell acquisition behavior in a hermit crab associated with different sensory cues. Hermit crabs are found from the tropics to the poles and most intertidal habitats have at least one species (reviewed in *Billock* 2008). Any visitor to a pet store will also know that there are terrestrial as well as marine hermit crabs. They are interesting because they have given up most of the 'armor' on the rear part of their body. This leaves their abdomens soft and vulnerable.

Some species seek protection inside bamboo or hollow mangrove roots. Others have adaptations to use coral, stones or wood for protection but most use discarded gastropod shells as a 'house'. Many species live in environments where appropriate shells are a limiting resource.

Research has shown that the quality of the gastropod shell is important to the survival and the reproductive success of hermit crabs. Most hermit crabs are quite selective about their shells. A combination of features determine shell quality including species from which the shell originates, size, weight, internal volume, size of the shell opening, color and shell condition (reviewed in *Billock* 2008). The shell cannot be too large or it is difficult for the animal to move, too light and it leaves the animal more vulnerable, particularly shell crushing predators. In some species males with heavier shells were demonstrated to be better able to guard females and so had better reproductive success (Yoshino, et al. 2004). A shell that is too light might also let the hermit crab be washed away by tidal currents. Pechenik & Lewis (2000) have shown that some species will reject snail shells with holes drilled in them by predators. Shell color affects visibility to predators and how much of the sun's energy is absorbed or reflected. A complicating factor is that as the animal grows the shell does not grow with it, so an animal will periodically need a new shell. Hermit crabs compete for available shells and there are complex intraspecific interactions involved. Many aspects of a crab's ability to survive are dependent on being able to find, acquire and keep a good shell.

Hermit crab behavior has been extensively studied. Locating, assessing and acquiring a shell requires the ability to decode a complex set of sensory cues and respond with the appropriate behavior. Some species use mostly visual cues from their **visual receptors** (complex compound eyes) to find new shells (reviewed in *Billock* 2008). Other species use primarily chemical cues, such as the scent of dying gastropods, dying members of their own species, the scent of predators or the concentration of calcium ions. Most hermit crabs use a combination of visual, chemical and tactile cues to locate a potentially useful shell. When they locate shells they continue to assess them using a combination of visual, chemical and tactile cues. A hermit crab will assess the size of the shell by grabbing it with its walking legs and then running its large front claws (chelae) over the surface. It will then look for the opening, rolling over the shell if need be. Once the opening is found the chelae are inserted, like a kind of measuring stick. If the animal wants the shell is sticks its abdomen into the shell and walks away (Mesce 1982).

Hermit crab **mechano-** and **chemoreceptors** are found on "hairlike" projections (sensillae) from the exoskeletons that are connected to nerves. Many sensillae respond to both chemical and mechanical stimuli. Crabs can have these sensillae distributed on many parts of their bodies. Often they are concentrated on the antenna and the appendages. Scanning electron micrograph studies found that the chelae of the Hairy hermit crab (*Pagurus hirsutiusculus*) had sensory structures that were very sensitive to calcium ions. So when the animal was running its chelae over the surface of a shell it was scraping it slightly to check the composition (Mesce 1993a, b). Mesce (1982) found that this species of hermit crab responded so strongly to calcium that it would choose plaster shells (very high calcium content) over real shells. In this lab you will observe hermit crab shell acquisition behavior and then you will design experiments to investigate further aspects of either hermit crabs' behavior or another species available in your lab.

Materials

- Hermit crabs (minimum 2 per group)
- Observation arenas (glass or plastic containers)
- Substrate material (sand or small-scale aquarium gravel)
- A range of shells: preferred species for your hermit crab type, alternate gastropod species, different sizes, shapes and colors
- Non-shell object of similar size and weight as home shell
- Shells of the preferred gastropod species that have been coated with a sealant a couple of days before (wax or aquarium silicone)
- Corks size appropriate for blocking shell apertures
- Plastic film
- Forceps
- Gloves
- Stop watch
- Balance for weighing shells
- Shell size piece of aragonite or calcite
- Sea water (28–30 ppt or concentration appropriate for hermit crab collection location)
- Additional materials for independent experiment to be discussed with instructor

Exercise 1
Learning to Recognize and Describe Behavior

When you begin to study the behavior of any animal, the first thing to do is to observe and take notes. This is sometimes referred to as *ad libitum* sampling.

1. Get into groups of 3-4 students. Collect a hermit crab and bring it back to your bench or table in a holding container, taking care to subject it to as few vibrations as possible. Remove the hermit crab **gently** from its shell. It is possible to do this carefully enough so as not to harm the animal. Place it back into its holding container.

2. Cover the bottom of the observation arena with a thin layer of substrate material (sand or small scale gravel). Wearing gloves place a group of shells into the observation arena. They should represent a range of species, sizes and colors. Weigh and measure the home shell so you can include at least one shell that is similar in size and type to the one the hermit crab is currently occupying. Distribute the shells evenly around the arena. If your hermit crab is aquatic, cover the shells with seawater.

Based on what you have read, generate a hypothesis about how the hermit crab will respond (i.e., can you make predictions about whether or not it will choose a new shell and if so which one).

Hypothesis:

3. Gently remove the hermit crab from the holding container and place it in the center of the observation arena. Observe the behavior of the animal for 4-5 minutes. Take notes and answer the questions.

4. At the end of the observation period, gently remove the animal and return it to the holding container on your bench top or table, with its home shell. Return to aquarium habitat or follow instructor directions. Animals should not have their shells removed for observations more than once in a lab period.

5. Pour out the seawater and rinse substrate and arena to prepare for the next exercise.

a) What did the hermit crab do when you placed it in the arena? Sometimes an animal needs a period of acclimation before it will respond to the cues in an environment. Did you see any evidence of this?

b) Did your results support your hypothesis or not?

c) Did it investigate any of the shells? If so, which ones?

d) Did the hermit crab crawl on top of the any of the shells and roll them or pick them up?

e) Did you observe scraping of the shell surface by the chelae? Did you observe insertion into the shell opening by the chelae?

f) Questions generated by your observations:

Exercise 2
Can a Hermit Crab Find its Shell Visually?

1. Remain in groups of 3-4 students. Collect a second hermit crab and remove the hermit crab **gently** from its shell. It is possible to do this carefully enough so as not to harm the animal. Place it into a holding container.

2. Wearing gloves, place a cork into the aperture of the home shell. Wrap the shell tightly in a single layer of thin plastic film. The characteristics of the shell must be visible but you are trying to block the chemical signals that might be coming from the shell. Wrap another non-shell object of similar size and color in plastic film as a control.

3. Place both objects in the observation arena at opposite ends or sides. Cover the shell and the control object with seawater.

4. Gently remove the hermit crab from the holding container and place it in the center of the observation arena. Observe the behavior of the animal for 4-5 minutes. Lengthen the time if initial observations suggested that the animal needed a period of acclimation. Take notes and answer the questions. Based on what you have read, generate a hypothesis about how the hermit crab will respond.

Hypothesis:

5. If the hermit crab orients to the home shell, check to see if it can 'track' the location of the shell visually. Move the shell around the arena with forceps. Record the response of the animal.

6. At the end of the observation period, gently remove the animal and return it to the holding container on your bench top or table. Pour out the seawater and rinse the arena and the substrate to prepare for the next exercise.

   a) Did the hermit crab locate the home shell? Did it appear to be able to 'see' the shell and recognize it? How would you know?

   b) Did your results support your hypothesis or not?

   c) The plastic film would have altered the tactile features of the shell and the chemical features of the shell. Did the hermit crab exhibit the behavioral repertoire associated with shell acquisition?

   d) It is difficult to isolate different kinds of stimuli. Describe an alternative experimental set-up to study just the effect of visual stimuli on hermit crab shell acquisition behavior.

   e) Questions generated by your observations:

Exercise 3
Can a Hermit Crab Find a Buried Shell?

1. Remain in groups of 3 - 4 students. You will use the same hermit crab from exercise 2, because it will still be out of its shell (think about the limitations of this approach).

2. Wearing gloves, remove the cork and the plastic film from the aperture of the home shell from exercise 2. Select a second shell of similar species, size, weight and color that has been coated with a sealant.

3. Empty the observation arena and rinse it out with seawater. Place enough sand in the bottom to cover the shells. Partially bury the shells at opposite sides of the arena (leave approximately 1 cm^2 shell area visible). Cover the arena with seawater. Based on what you have read, generate a hypothesis about how the hermit crab will respond.

Hypothesis:

4. Gently remove the hermit crab from the holding container and place it in the center of the observation arena. Observe the behavior of the animal for a minimum of 30 minutes. If after 30 minutes the animal has not located either shell, stop taking notes, gently remove the animal and return it to the holding container on your bench top or table. Place the home shell into the holding container with the animal.

   a) What stimuli would the hermit crab be responding to primarily if it oriented to the buried home shell?

   b) Did your results support your hypothesis or not?

   c) How would you determine the difference between an animal being attracted to something versus being repelled by something else (i.e., how would you make sure the sealant did not repel the crab)?

d) Describe an alternative experimental set-up to study the effect of chemical stimuli on hermit crab shell acquisition behavior.

e) How could you quantify your observations in exercises 1-3?

f) If you made conclusions about hermit crab behavior based on your observations from exercises 1-3, what are some of the limitations (problems)?

g) Questions generated by your observations:

Exercise 4
Designing Your Own Experiment

In groups of 3-4 you will now work to design your own experiment. Using Hermit crabs or an animal approved by your instructor, design an experiment to investigate the **primary sensory receptor** used by a marine organism to respond to cues in its environment. You have some pilot or baseline data on hermit crab shell acquisition behavior. You may choose to continue to investigate how crabs response to different stimuli when seeking shells (e.g., a block of aragonite or calcite (different forms of $CaCO_3$) is available to represent a chemical stimulus; or you can investigate the sensory responses to other important resources e.g., food; or you can investigate the response to members of the same species [social interactions] like competition for shells). If you have access to other animals in your lab and wish to use those, you will need to do some research about their biology to help you with your experimental design.

There are a number of possibilities for investigations of sensory reception of marine organisms. You need to take into account the kinds of sensory receptors present in your experimental organisms. General questions for possible experimentation are what are the behavioral responses to food, predators, and competition from other organisms.

1. Pick one of these questions, examine the list of available materials, and think about the **independent (experimental) variables you** can manipulate to provoke a behavioral response in your organism. For example, if your group decides to investigate food responses, how

would you present food cues? What would you use for a control? Does the animal use visual, tactile, and/or chemoreceptors to find the food? How would the type of sensory receptor affect your choice or choices of food stimuli used in the experiment?

Independent variable for question 1: _____

2. Now that you have selected your independent variable(s), look at the background information you have collected and pick a dependent variable (response variable) for your experiment. The **dependent variable** needs to be something you can observe, count, or measure.

Dependent variable _____

3. You need to discuss your choice of experimental variables with your lab instructor, who will help guide you by asking you questions to clarify any points that you may not have considered before you make your final choice. Your lab instructor will approve your variable choices and initial your lab manual before you can begin designing your experiment.

Instructor approval _____

4. Write a hypothesis that describes the relationship between the independent and dependent variables you have chosen using the "if/then" format. For example, if you were interested in investigating the relationship between the amount of dissolved oxygen in the water and the rate at which fish die, you could write your research hypothesis as follows:

   As the amount of dissolved oxygen in the water decreases, fish mortality increases.
   In this example, dissolved oxygen concentration is the independent variable and fish mortality is the dependent variable. The hypothesis describes the relationship between these two variables and makes a prediction about the outcome of the experiment.

   Write your experimental hypothesis in the space below and have your lab instructor approve it before designing your experiment.

Hypothesis:

Instructor approval _____

Experimental Design Preparation

Now you that you have identified your variables of interest and made a prediction about the relationship between these variables, you are ready to begin thinking about how you will conduct your experiment. You will need to make observations (*ad libitum* sampling) of your experimental organism for a brief period of time prior to starting your experiment, so that you will be able to describe baseline behavior for your organism in the absence of any experimental cues or stimuli. If you continue investigating hermit crab shell acquisition behavior you may use the baseline data collected in exercises 1-3 with your instructor's permission. Answering the following questions as a group will assist you with outlining your experimental procedure. You will have about 10 minutes to discuss the questions below and develop your experimental protocol.

1. How many and which organisms do you plan to use for your experiment?

2. Will you repeat the experiment? If so, how many times?

3. Will you use the same animals for each experiment (repeated trials)?

4. Do you need to wait a while before repeating the experiment using the same animals? If so, why and how long would you wait between trials (repeats)?

5. Do you need to change the water in your test container between experimental trials? If yes, why (how would it impact your experiment if you used the same water for all trails)?

6. Do you have a <u>control</u>? What would you use as a control?

Designing your Experiment

In the space below, construct an outline for your experimental protocol and number your steps. Be as specific as you can and include details about the equipment used, number of organisms used, and type and frequency of stimulus presentation, for example. Your lab instructor must approve your experimental procedure before you can begin.

Instructor approval _____

Results

1. Record your observations in this space. Include any notes about the behavior of your organism after the acclimation period but prior to the start of the experiment. By way of example, if you are measuring the amount of time it took for your organism to respond to the presence of a potential predator, your dependent variable would be time and the units for that variable would be some time interval of your choosing, such as seconds or minutes. Do not forget to include observations you made with the control.

Observations:

2. Design a data table (below) to summarize your data, with columns for the independent and dependent variables and rows for each organism or trial. Also, do not forget to include the units on your column headers

3. Summarize your data visually by constructing a graph on the paper provided. Discuss with your lab partners which type of graph (histogram=bar chart, line graph, or scatter plot) is most appropriate for the kind of data you collected. Be certain that you label the horizontal axis (x-axis) with your independent variable name and include units, and label the vertical axis (y-axis) with your dependent variable name including units.

Discussion and Conclusions

Reviewing your data and analyzing your graph, was your original research hypothesis supported? Summarize your findings in the space below.

What recommendations would you make to students who would like to repeat your experiment? In other words, what would you do differently next time?

What was your "take home message" for this experiment?

Describe one interesting fact or outcome that you did not expect.

Questions

You will need to write a lab report summarizing your group's experiment and submit your summary to your instructor. Points to include in your summary:

1. An introduction including the common name and scientific name of your experimental organism, your independent variable(s) and your dependent variable, your research hypothesis.

3. Brief description of your experiment, including methods of testing.

4. Results section where data are presented.

5. Conclusion including how your findings compare to your hypothesis, and what other conclusions you think are implied by your results.

Suggested Reading

Billock, W. L. 2008. Evidence for 'Contexual decision hierarchies in the hermit crab, *Pagurus samuelis*. (Doctoral Dissertation) 155 pp. Retrieved from ProQuest Dissertations and Theses. (Accession Order No. 3333191).

Krebs, J.R., Davies, N.B.,and S. A. West. 2012. An Introduction to Behavioural Ecology, Fourth Edition. Wiley-Blackwell Publishing, Oxford, England, 520 pp.

Mesce, K. A. 1982. Calcium-bearing objects elicit shell selection behavior in a hermit crab. Science 215: 993-995.

Mesce, K. A. 1993a. Morphological and Physiological Identification of Chelar Sensory Structures in the Hermit Crab *Pagurus hirsutiusculus* (Decapoda). Journal of Crustacean Biology, 13:95-110.

Mesce, K.A., 1993b. The shell selection behavior of two closely related hermit crabs. Animal Behaviour 45, 659-671.

Pechenik, J.A. and S. Lewis, 2000. Avoidance of drilled gastropod shells by the hermit crab *Pagurus longicarpus* at Nahant, Massachusetts. J. Exp. Mar. Biol. Ecol. 253, 17-32.
Persons, M.H., Uetz, G.W., 1996.

Yoshina Yoshino, K., Ozawa, M., Goshima, S., 2004. Effects of shell size fit on the efficacy of mate guarding behaviour in male hermit crabs. J. Mar. Biol. Assoc. U.K. 84, 1203-1208.

Glossary of Terms

ad libitum **sampling:** " Ad lib", casual sampling where the observer watches and gathers as much data as possible to provide a baseline overview of the behavior of the animal.

chemoreceptor: A sensory receptor that transmits information about the total solute concentration in a solution or about individual kinds of molecules.

chemotactic response: A response or change in animal orientation with the presentation of chemical cues.

control: The portion of the experimental population that does not receive the treatment, for this group all variables are held constant. Results from the experimental groups are compared for each variable with the measurements from the control group for those variables, to determine if differences exist as a result of the experimental "treatment."

dependent variable: The variable or factor in an experiment whose values change in response to the manipulation of the independent variable.

Heterotrophy: (*Hetero*, different; *troph*, nutrition): Organisms that can not manufacture their own food and must ingest organic carbon and nitrogen produced by other organisms.

independent variable: A variable or factor that the experimenter manipulates (changes).

mechanoreceptor: A sensory receptor that detects changes in shape or physical deformation in the body's environment due to pressure, touch, motion, stretch, and sound.

variable: A characteristic in an experiment that assumes more than one value; for example, temperature is a variable if measurements are made at different temperatures, such as 0 degrees C, 10 degrees C, 25 degrees C, etc.

visual receptor: A sensory receptor that detects light and/or movement; examples include eyespots, ocelli, and "camera eyes" (like those in a squid).

Lab 12

Marine Mammal and Reptile Movements

Objectives

Upon the completion of this exercise, you should be able to:

1) Describe the general seasonal pattern of North Atlantic Right whale migration.

2) Describe the conservation efforts to reduce **ship-strike** for the North Atlantic Right Whale.

3) Describe the migratory patterns of sea turtles and how feeding locations differ from breeding and nesting locations.

4) Given latitude and longitude coordinates, be able to plot the travels of a marine mammal or reptile on a map.

5) Calculate the daily distance traveled for a marine mammal or reptile from the latitude and longitude coordinates provided in the online data set.

6) Summarize your findings in writing and provide an explanation for the rate and direction of movement for an organism of your choice, given what you have learned about the life history of marine mammals and turtles.

Background: Whales

The largest group of marine mammals is the Cetaceans, which includes whales, dolphins, and porpoises. These charismatic megafauna have fascinated mankind for centuries. Streamlined bodies insulated with thick blubber are uniquely adapted for life in the ocean. Cetaceans are divided into two distinct groups: the suborder Mysticeti, the baleen whales, and the suborder Odonticeti, the toothed whales.

Populations of whales can inhabit a very large area and travel long distances. The Pacific Gray Whale (*Eschrichtius robustus*) has one of the longest migrations routes known for any mammal, 8,000–11,000 kilometers (5,000–6,800 mi) each way. Whales generally begin the migration toward warmer tropical waters in late summer, so that they reach breeding and calving grounds in warmer tropical waters. The longer movements (macro-movements) are usually called migrations, and the endpoints of the migrations are places where they feed and where they breed. The feeding areas are usually toward the polar waters where the waters are more productive, and the breeding areas are usually in tropical or subtropical waters where the waters are warmer, but less productive.

The shorter movements (micro-movements) are how whales use the local habitat in which they are located. These localized movements are usually foraging movements in feeding areas, but in warmer breeding areas there are still many questions to be answered. In this lab we will focus on how an understanding of the movements of the endangered North Atlantic Right Whale is important to conservation efforts.

The North Atlantic Right Whale (*Eubalaena glacialis*) is one of three species of right whale (Figure 12-1). It is named because whalers as early as the 1500's thought it was the 'right' whale to hunt. It is a large baleen whale that grows to 13 to 18 m (43 to 59 feet) in length and can weigh up to 55 metric tons (60 tons). It has few natural predators-only the largest sharks and the Killer Whales (*Orca*). But it is a very slow swimmer (less than 10 knots) and floats when it dies. It also spends much of its time in shallow water fairly close to the shore. This made harvesting it for valuable blubber much easier than for other species of whales and as a result it was hunted almost to extinction. The dire predicament of this species was recognized and in 1935 a ban on hunting was declared. However, the number of animals left in the wild was very low and the average female only breeds six times so that despite the reduction in whaling pressure the population did not rebound quickly. In 2012, 77 years after the ban on hunting, less than 400 North Atlantic Right Whales exist and the population remains critically endangered.

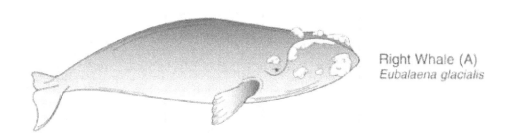

Right Whale (A)
Eubalaena glacialis

Figure 12-1 North Atlantic Right Whale

The two most significant sources of mortality for North Atlantic Right Whales are ship-strikes and entanglement in fishing gear. Collisions with ships were responsible for 24 of the 67 reported deaths between 1970 and 2007. This is probably an underestimate because many strikes are not noticed or reported. The natural habitat and migratory patterns of the North Atlantic Right Whales put them directly in the major shipping lanes off the East Coast of the United States-some of the busiest commercial waters in the world. Other aspects of their biology also make them particularly vulnerable. They feed by skimming; opening their mouths and collecting zooplankton as they swim along, often at the surface. Although they communicate by sound and have good hearing they do not appear to respond to the presence of ships. They will not move out of the way of an oncoming ship, especially when the whales are gathered in large groups (5-30 animals) to carry out courtship behavior. They are darkly colored and lack a dorsal fin so they can be difficult for mariners to see. Since the population is so small preventing even a few deaths could be significant for recovery so **NOAA's** National Marine Fisheries Service is coordinating a large conservation effort to reduce ship-strikes.

Conservation efforts are focused on education, identifying critical habitat and regulating shipping in those areas, as well as providing accurate, almost real-time whale location information as an early warning system. The coastal waters of the South Eastern United States (Florida and Georgia) are where the animals calve. They migrate north to feeding grounds off the coast of New England, Nova Scotia and in the Bay of Fundy. Other critical areas include the Great South Channel, Massachusetts Bay and Cape Cod Bay. The migratory path takes the whales past the entrance of most of the major shipping ports on the eastern seaboard from Savannah, GA to Providence, RI. The regulations affecting shipping reflect the dynamic location of the whales. There are **seasonal management areas** (SMA). These are areas where the whales are consistently present at certain times of year. There are also **dynamic management areas** (DMA), these can be established temporarily, when a certain number of whales are known to be in the area. All vessels 19.8m (65 ft) or greater are required to reduce their speed to 10 knots or less in an SMA and they are asked to do the same when they enter a DMA. At certain times of year there are also regions where the whales are known to congregate that ships 300 tons and larger are asked to avoid. Ships this size must report in when they are entering designated areas **(Right Whale Mandatory Reporting System)**. A good system for keeping track of where the whales are and communicating this information to shipping traffic is key to making this conservation effort work. Whale watch stations are set up on land, aerial surveys are flown and a series of buoys equipped with underwater microphones have been set up along one of the major routes into Boston Harbor. Ships can receive up-to-date information many different ways, including through regular navigation communication, updated navigation channels and just recently through a Whale Alert (EarthNC) app that incorporates up-to-date maps of right whale management areas, required reporting areas and recommended routes with near real-time-Boston-area acoustic whale detection buoys.

In this lab you will use Right Whale sighting data and a map to try to determine the direction of migration and areas where shipping traffic should be regulated, and at what time of year a particular area would be expecting whales and might be designated a Seasonal Management Area.

Materials

- Pencil
- Calculator
- Atlas (or internet access)
- Internet access
- iPad®/iPhone® (optional)
- Ruler

Exercise 1
North Atlantic Right Whale

The range of the North Atlantic Right Whale extends primarily from Florida to Nova Scotia. In this portion of the lab, you are going to plot North Atlantic Right Whale sightings on the map and decide at what **time of year** you would make a recommendation to a regulatory body that an area known to be critical habitat might be designated a Seasonal Management Area.

1. Use an atlas or the internet to locate the following five areas known to be critical habitat:

     a. Coastal Florida and Georgia

     b. Great South Channel

     c. Massachusetts Bay and Cape Cod Bay

     d. Bay of Fundy and Gulf of Maine

     e. Scotian Shelf

Label those areas on the map of Eastern North America provided in Figure 12-2.

2. Use the sighting data for Whale 2420 provided in Table 12-1. Plot these data on the map provided (Figure 12-2), use a small arrow to indicate the direction of movement and include the date.

3. Plot the movements of at least one other individual. Using data from The New England Aquarium's special section: The Right Whale Catalog http://rwcatalog.neaq.org/, for a

comprehensive data base of all identified North Atlantic Right Whales). Choose an individual with minimum of 3-4 years of data.

4. Use the plot of the individual whale movements and the summary data in Table 12-2 to suggest a time frame for each critical habitat to be a Seasonal Management Area.

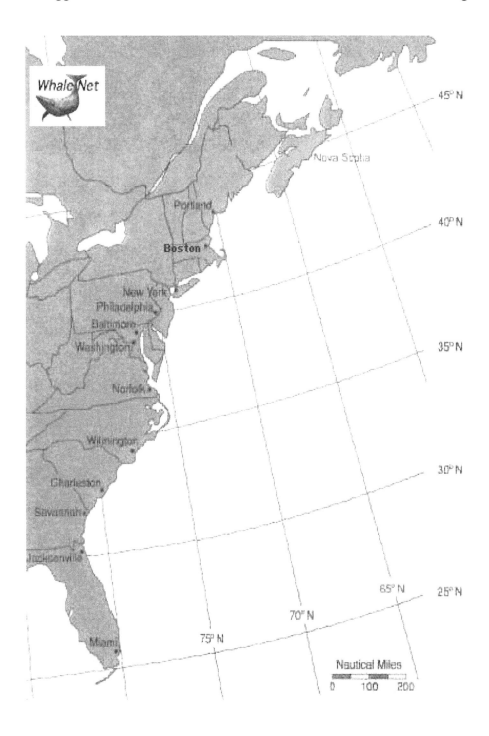

Figure 12-2 Map of Eastern North America with Latitude and Longitude

Table 12-1 Location information for Whale 2420 (from The Right Whale Catalog)

Whale ID 2420	Female
1/5/1994	Georgia
1/7/1994	Florida
1/12/1994	Florida
1/25/1994	Florida
1/17/2002	Georgia
1/18/2002	Georgia
7/2/2003	Gulf of Maine
6/9/2005	Great South Channel
2/10/2006	Georgia
2/22/2006	Florida
2/27/2006	Georgia
3/3/2006	Georgia
12/20/2010	Florida
12/25/2010	Florida
12/28/2010	Florida
12/30/2010	Florida

Table 12-2 Total Number of right whale sighting reports to the Right Whale Sighting Advisory System in 2011 (Khan et al. 2012)

Region	Jan	Feb	Mar	Apr	May	Jun	Jul	Aug	Sept	Oct	Nov	Dec
NE (ME through NY)	72	52	130	187	106	27	7	16	31	20	64	16
Mid-Atlantic (NJ through VA)	2	4	5			2						3
Canada							1	14	22		3	

5. Fill in the time period for each critical habitat area that the whale sighting data suggests is relevant:

        a. Coastal Florida and Georgia from _____ to _____

        b. Great South Channel _____ to _____

        c. Massachusetts Bay and Cape Cod Bay _____ to _____

        d. Bay of Fundy and Gulf of Maine _____ to _____

        e. Scotian Shelf _____ to _____

6) The coastal waters of the southeastern U.S are where the North Atlantic Right Whales give birth. The aerial survey data from Dec 2010 to early March 2011, sorted so that all the sightings

of a particular calf are grouped together, are provided in Table 12-3. Choose 5 individuals and plot their movements on Figure 12-3. Based on this subsample, shade in the area the data suggest should be designated a Seasonal Management Area.

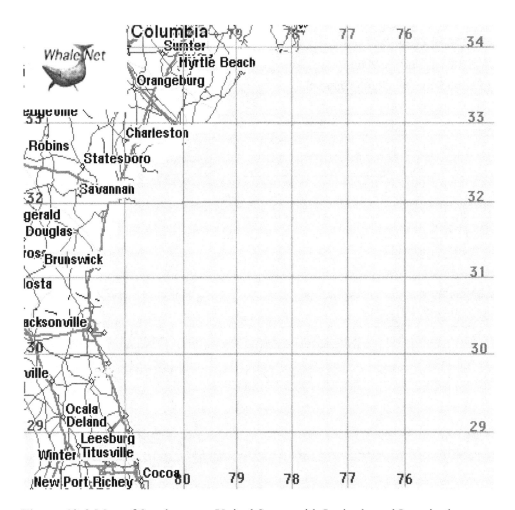

Figure 12-3 Map of Southeastern United States with Latitude and Longitude

Table 12-3 2010/2011 Early Warning System Aerial Survey Data for North Atlantic Right Whale Calves (from Jackson et al. 2011)

Aerial Survey Data	Latitude	Longitude	Whale ID number
12/25/2010	30.64633	-81.22701	2011CalfOf1243
12/30/2010	30.5775	-81.25617	2011CalfOf1243
12/31/2010	30.58134	-81.26684	2011CalfOf1243
1/4/2011	30.523	-81.28817	2011CalfOf1243
2/13/2011	30.44333	-80.83549	2011CalfOf1245
2/24/2011	30.78	-81.11684	2011CalfOf1245
3/5/2011	30.6935	-81.09267	2011CalfOf1245
12/18/2010	30.7388	-81.3048	2011CalfOf2029
12/20/2010	30.47667	-81.24216	2011CalfOf2029
12/25/2010	30.5685	-81.30067	2011CalfOf2029
12/25/2010	30.56667	-81.301	2011CalfOf2029
12/30/2010	30.40117	-81.27116	2011CalfOf2029
12/30/2010	30.4	-81.28467	2011CalfOf2029
1/15/2011	30.339	-81.1205	2011CalfOf2029
1/29/2011	30.473	-81.03783	2011CalfOf2029
1/29/2011	30.47117	-81.04601	2011CalfOf2029
2/21/2011	30.62234	-81.1755	2011CalfOf2029
1/20/2011	30.47217	-81.05617	2011CalfOf2040
2/21/2011	30.41683	-80.97701	2011CalfOf2040
2/28/2011	30.32917	-81.24067	2011CalfOf2413
12/20/2010	30.53267	-81.19534	2011CalfOf2420
12/25/2010	30.3985	-81.16167	2011CalfOf2420
12/28/2010	30.5735	-81.17033	2011CalfOf2420
12/30/2010	30.6145	-81.18383	2011CalfOf2420
12/30/2010	30.61133	-81.18884	2011CalfOf2420
1/29/2011	30.39117	-81.011	2011CalfOf2420
2/16/2011	30.44333	-81.37683	2011CalfOf2420
12/25/2010	30.54133	-81.31184	2011CalfOf2746
1/23/2011	30.42867	-80.89633	2011CalfOf2746
12/30/2010	29.95417	-81.27451	2011CalfOf3010
12/31/2010	30.24384	-81.21017	2011CalfOf3010
1/30/2011	30.46783	-81.25383	2011CalfOf3010
2/17/2011	30.28983	-81.1095	2011CalfOf3270
2/24/2011	30.49567	-81.12566	2011CalfOf3270
2/17/2011	30.37634	-81.11684	2011CalfOf3293
1/4/2011	29.938	-81.2545	2011CalfOf3430

Questions

1. When you followed the movement patterns of individual whales for a number of years was there a there a pattern, i.e., how would you figure out if the whale was migrating?

2. Why are the sightings so much more common in some places compared to others? Does this always reflect the biology of the animal?

3. What are some of the difficulties that a regulatory agency might face when trying to decide where to put the boundaries of a conservation area?

4. Was it clear when to designate the time frame for the Seasonal Management Areas?

5. **Optional activity**. Download the Whale Alert app to an iPad® or an iPhone® and use the maps of the managed areas to check the location of Seasonal Management Areas, Mandatory Reporting Areas. Check to see if the acoustic buoys have picked up any whale activity recently. (At the time of writing this app is only for iPad® and iPhone® platforms, but in future it might be available for Android® phones as well.)

Turtles

Sea turtles are largely confined to warmer tropical and subtropical coastal areas, in contrast to the great whales, which can be found in both coastal and open ocean areas. There are seven species of sea turtles: leatherback, hawksbill, Kemp's Ridley, green, loggerhead, flatback, and the Olive Ridley. Some sea turtles, the loggerhead for example, are generalists, eating a highly varied diet. Other sea turtles have highly specialized diets. The hawksbill sea turtle is commonly found on reefs in tropical areas where large sponges, its preferred food, are common (Figure 12-4).

Each species of sea turtle has specific nesting sites located in warmer temperate or tropical coastal areas. Female turtles come on shore to nest during the night, digging a shallow pit in the sand into which they lay their eggs. A female will lay between 80–150 eggs before covering the nest and returning to the sea. The sex of an incubating turtle is determined by nest temperature. When the nest is warmer than 29.9°C, the embryos will be female. At temperatures below 29.9°C, the embryos will be male.

The eggs hatch at night, and the young turtles move toward the brightest point on the horizon, usually the ocean illuminated by the moon. This is why many states require that beach residents turn off outside lights during turtle hatching season, preventing the lights from causing young hatchlings to move in the wrong direction. Each turtle species nests on a few select beaches, usually in remote areas with little coastal development. Like the great whales, turtles will migrate from coastal feeding areas to breeding areas as far as 1400 miles away. Tagging studies on the green sea turtle indicate that a large population that feeds on manatee grass along the Brazilian coast migrates to Ascension Island, an island in the south Atlantic between Africa and South America, to mate and lay eggs.

In this lab exercise, you will visit one of several internet sites that provide students with access to satellite tagging data and follow one or more marine animals (whales, dolphins, porpoises, seals, sea lions, or sea turtles) using data collected over several months to a year, to make observations about your animal's movement.

Sea Turtle Species	Food Preferance
Leatherback 	Jellyfish
Hawksbill 	Sponges
Kemp's Ridley 	Crabs, Shrimp, snails, clams, jellyfish, sea stars, and fish
Green 	Turtlegrass and manateegrass
Loggerhead 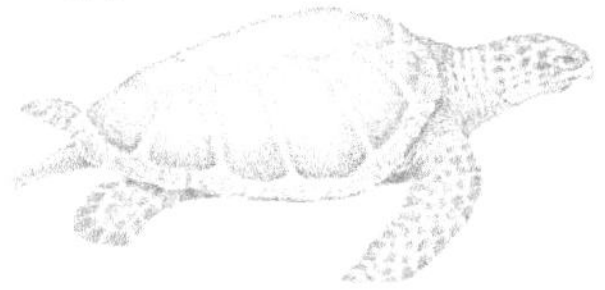	Conchs, clams, crabs, horseshoe crabs, shrimp, sea urchins, sponges, fish
Flatback 	Sea cucumbers, jellyfish, mollusks, prawns, bryozoans, and seaweed
Olive Ridley 	Jellyfish, snails, shrimp, and crabs

Figure 12-4 Food Preferences of Sea Turtle Species

Calculating Time, Distance, and Speed

When looking at tracking data you need to make some calculations so you can evaluate the path of an animal. To calculate time between sightings, remember that recorded sighting times use military time. Any time point greater than 12 hours is in the second half of the day. For example, if the initial sighting is 9:00 on Monday and the second sighting is 14:00 on Friday, you would use the following equation: 24-9 = 15 hours elapsed for Monday; Tuesday through Thursday 3 x 24 hours elapsed = 72 hours elapsed; and then 14 hours elapsed for Friday. Sum these values as follows: 15 + 72 + 14 = 101 hours elapsed between sightings.

To calculate distance between the two sightings, use the formula of the distance derived from the Pythagorean Theorem:

$$d = \sqrt{(x_2 - x_1)^2 + (y_2 - y_1)^2}$$

Where:

d – distance (in degrees)

x_2 – degrees Latitude for second coordinate

x_1 – degrees Latitude for first coordinate

y_2 – degrees Longitude for second coordinate

y_1 – degrees Longitude for first coordinate

(note if the animal crossed the equator or the prime meridian this formula will not be accurate)

If Monday's 9:00 coordinate is 31.05 N 81.75 W, and Friday's 14:00 is 30.46 N 81.11W, then the distance between two points is the square root of [$(30.46-31.05)^2 + (81.11-81.75)^2$] = square root of [$(-0.59)^2 + (-0.68)^2$] = square root of [0.3481 + 0.4626] = square root of 0.8105 = 0.900278 degrees. The distance in kilometers from Table 12-3 is roughly 95.997 km (halfway between 30° and 31°).

To calculate distance traveled by the organism, multiply 95.997km/° by the value above, 0.900278° = 86.42 km. To calculate the speed of movement, divide the distance traveled, 86.42 km, by the time elapsed, 101 hours, and the rate of movement for this tagged animal is 0.847 km/hour. All of this assumes a linear path for the animal, so to whatever extent the actual path curves or doubles back on itself, both distance and speed calculated in this way will be underestimates.

Exercise 2
Loggerhead Sea Turtles

Loggerhead sea turtles are common on the Atlantic coast of Georgia. Loggerheads have nesting sites on many of Georgia's less-developed coastal islands. Let's look at several loggerhead sea turtles that were fitted with satellite **tags** and released off the coast. You will calculate the distance the turtles traveled between satellite observations, as well as calculate an average rate of movement (distance/day). In addition, you will use the map provided and plot the movement of three turtles.

1. Using the data for "Annie in Table 12-4 (from http://whale.wheelock.edu/whalenet-stuff/loggerhead_data1.html). Calculate distance traveled and rate of travel. Remember that to calculate time you need to figure the elapsed time (how much time has passed) between satellite observations and record that value in minutes (remembering that 60 minutes = 1 hour).

2. You will need to use latitude and longitude to calculate distance, and there are a couple of values you will need to know in order to do this. Use the longitude coordinates to determine east-west distance by subtracting the largest number from the smallest number and the formula given in the introduction. Use the adjusted longitude data that is expressed in decimals.

3. You will need to compute how many degrees the turtle traveled, and using the latitudinal data above, compute distance traveled in kilometers in the data table provided.

4. You will need a ruler to measure Annie's path on your map (Figure 12-5) and determine the scale by measuring the distance on the ruler between 1° latitude, and use the relationships in Table 12-5. Calculate the rate of movement by dividing the distance (km) by elapsed time (hours).

5. Repeat the steps above to plot the movements of the other two loggerhead turtles: "Isabelle" and "Aeriel." Plot the dates and connect the points with an arrow indicating direction of movement. To access the data go to http://whale.wheelock.edu/whalenet-stuff/loggerhead_data1.html.

12-4 Movement Data Summary for Loggerhead Sea Turtle "Annie" (Williamson 2005) Latitude and Longitude expressed in degrees and decimals.

Date	Time (hours)	Latitude	Adjusted Longitude	Difference (=distance, km)	Rate (distance/time in km/hr)
6/15/97	13:29:40	32.481	80.248		
6/15/97	23:05:15	32.482	80.367		
6/17/97	19:34:45	32.535	79.87		
6/21/97	9:06:23	33.388	79.131		
6/21/97	13:01:40	33.352	79.163		
7/1/97	12:39:37	32.919	79.525		
7/6/97	06:25:11	32.924	79.531		
7/8/97	07:40:56	32.797	79.75		
7/9/97	07:29:58	32.787	79.767		

Table 12-5 Distance in Kilometers per Degree (°)

Latitude	Distance $1°$ Longitude (kilometers, km)
$30°$	96.488
$31°$	95.506
$32°$	94.495
$33°$	93.455
$34°$	92.347

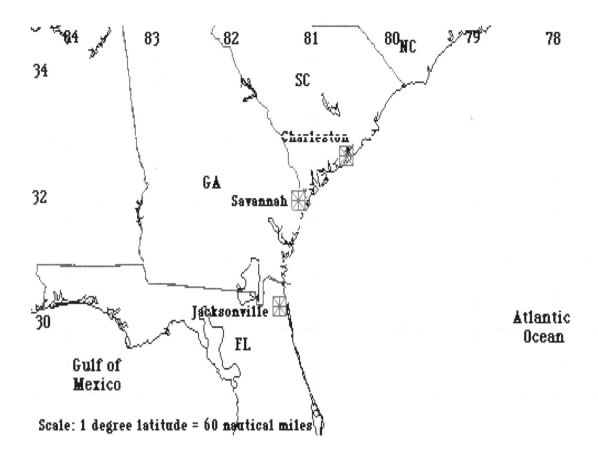

Scale: 1 degree latitude = 60 nautical miles

Figure 12-5 Map of the Southeastern United States used to Plot Loggerhead and Sea Turtle Movements

Questions

1. Are the loggerheads traveling a route that you expected? Why?

2. What is the normal habitat for a loggerhead turtle?

3. What is the animal's relative (average) rate of travel?

4. Do you think the estimates of distance traveled and rate of travel were accurate?

5. Discuss some of the possible sources of error when working with satellite tracking data. http://whale.wheelock.edu/whalenet-stuff/sat_tags_work.html is a website you might want to visit before answering this question.

Exercise 3

Choose Your Own Animal

Now it is your turn to choose an animal to track and follow. Visit the WhaleNet website: http://whale.wheelock.edu/whalenet-stuff/stop_cover_archive.html. This site contains archived satellite tagging data for many marine animals. You can select any animal that you would like, as long as there are at least three to four months of data for the organism. Check the dates in the database to ensure you have enough data to plot and make statements about your animal's movements. Animals in the database that are commonly tracked in the southeastern U.S. and the Florida panhandle include dolphins, Right Whales, Kemp's Ridley sea turtles, and loggerhead sea turtles. Your instructor will give you a due date.

1. Create a data table similar to Table 12-4. You can use Microsoft Excel or create a table in a word processing application.

2. Create a tracking map for your organism. Bitmap and jpeg files for mapping are frequently included with summary data. You will need to save the map as a separate file and then copy the map into your word processing file. Then, plot the path of your animal on this map. Be certain to include the URL and name of the internet source of your map in your written summary.

3. Summarize what you know about the life history of your organism, including the following: average life span, age of first reproduction for males and females, average number of offspring, and seasonal migration patterns (i.e., Where does it feed, and where does it go to reproduce?). You should include any behaviors that might explain the movements of your organism.

   Discuss the movements of your animal and explain why it went where it did using what you have discovered about its life history. Do you have any information about the age or sex of your animal that might be useful when interpreting its movements?

Suggested Reading and References

Khan C, Cole T, Duley P, Henry A, Gatzke J, Corkeron P. 2012. North Atlantic Right

Whale Sighting Survey (NARWSS) and Right Whale Sighting Advisory System (RWSAS)

2011 Results Summary. US Dept Commer, Northeast Fish Sci Cent Ref Doc.

12-09; 6 p. Available from: National Marine Fisheries Service, 166 Water Street, Woods

Hole, MA 02543-1026, or online at http://www.nefsc.noaa.gov/nefsc/publications

Jackson, K. A., J.L. Jakush and J. G. Ortega-Ortiz. 2011.

Aerial Surveys for Ship Strike Mitigation and Other Field Observations of North Atlantic

Right Whales (*Eubalaena glacialis*) off the East Coast of Florida and Georgia

December 2010-March 2011. Central Early Warning System

Florida Fish and Wildlife Conservation Commission (FWC)

North Atlantic Right Whale

NOAA. 2012. North Atlantic Right Whale Sighting Survey and Sighting Advisory System,

http://www.nefsc.noaa.gov/psb/surveys/

 Williamson, M. J. 2005. WhaleNet. /http://whale.wheelock.edu/whalenet-

stuff/loggerhead_data1.html.

Glossary of Terms

Dynamic management areas (DMA): If an aggregation of Right Whales suddenly or
unexpectedly appears in an area a DMA can be designated and shipping traffic in that area is
asked to voluntarily reduce speed.

NOAA: National Oceanic and Atmospheric Administration is the federal agency focused on the
oceans and the atmosphere.

Right Whale Mandatory Reporting system: Ships greater than 300 tons are required to report
in to a shore-based station when they enter critical whale habitats.

Seasonal Management Areas (SMA): Areas where Right Whales are regularly found at certain times of year in relatively large numbers. During the dates in which an SMA in is effect, ships are subject to regulations such as moving at a reduced speed.

ship-strike: Occurs when a vessel hits an animal, this is the most important cause of mortality in North Atlantic Right Whales.

tags: Instruments that send signals to satellites maintained by the ARGOS system in Largo, Maryland, and Toulouse, France. This information is used to locate both fixed and movable positions on the Earth's surface.

Acknowledgement: *This exercise is an extension of Dr.Patricia Yager's Marine Sciences Department, University of Georgia, environmental monitoring projects in her Honors Introductory Marine Biology course. Thank you for helping connect inland students to the activities of coastal marine mammals and reptiles.*

Lab 13

Intertidal Communities

Objectives

Upon completion of this exercise, you should be able to:

1) Distinguish between the **abiotic** and **biotic** factors in an intertidal zone.

2) Describe adaptations needed to survive in an **intertidal (littoral) zone** by specific organisms.

3) Describe the different types of intertidal zones (**rocky, muddy and sandy**).

4) Observe differences between physical and chemical parameters that exist in intertidal zones.

5) Be able to identify organisms of the intertidal **ecosystems**.

Intertidal Zones

The **intertidal zone** is the narrow fringe along the shoreline that connects the oceanic **ecosystem** with the **terrestrial** ecosystem. It is the area between high and low tide, sometimes referred to as the **littoral zone**. The organisms that are able to survive in the littoral zone are subjected to many stressors as they are alternately exposed to the atmosphere and seawater through the action of tides. The variety in conditions between different parts of the intertidal zone can result in great diversity in intertidal communities.

Intertidal communities are often characterized by their substrate type. Rocky shorelines are characterized by steep slopes and a rocky substrate. They can be found along **active continental margins** (e.g., much of the western Americas), regions where glacial action has scraped all the sediment from the continental shelf, areas subjected to strong wave actions, active lava flows or at the site of ancient coral beds. Alternatively **passive continental margins** favor more accumulation of sediments. This results in a **soft-bottom** intertidal shoreline, either **sandy** or **muddy**. Most of the east coast of North America and Gulf of Mexico are considered to be soft-bottom shorelines. The sediment source and level of water motion determine the characteristics of the community.

In this lab we will discuss the physical and chemical parameters, and practice measuring them as well. The chemical and physical factors are the **abiotic**, or nonliving, characteristics of the **ecosystem**. **Biotic**, or living, parameters of an ecosystem include organismal adaptations and vegetative adaptations. In order to understand the differences between the abiotic and biotic factors of the intertidal ecosystem, we will do several field measurements. First it is important to understand the ecosystem and the challenges encountered by the organisms that inhabit the intertidal zone.

Adaptations in the Intertidal Environment

Intertidal zones are a unique environment, and the organisms that reside there are faced with several challenges. Some of these challenges include desiccation, temperature extremes, exposure to UV radiation, the pounding of the waves, varying salinity, and strong currents

Organisms that live in this zone, such as mussels, sea urchins, snails, and anemones, must be adapted to both wet and dry conditions. Mussels and other shell organisms close their shells to keep moisture in. Sea urchins carve holes in rocks and anemones will retract their tentacles to store moisture during dry conditions. Other organisms such as snails secrete mucus. Some like the algae, sea lettuce (*Ulva*) are very tolerant to desiccation. Many of these adaptations also reduce temperature change. Organisms that can move often find a place to hide during low tide. A good place to look for small crabs and isopods during low tide is under a mat of seaweed. A light shell color can also reflect solar radiation. Some sea anemones will attach small pieces of light shell to their pedal stalks as a protective measure.

Adaptations to wave action include flexibility like some types of seaweed (e.g., Sea Palms, *Postelsia*) or having a very low profile. Limpets found on an open rocky coast tend to have a lower profile than those found in a more protected bay. Organisms like chitons have a dorsally flattened body in a shape that helps dissipate the wave force and they also have the ability to use their fleshy foot to attach to a rocky surface. To avoid being carried away by a tidal currents organisms such as sea stars and anemones can latch on to rocks. Mussels use their byssal gland to secrete byssal threads to attach to the rocks. Barnacles cement themselves to the surface. Many organisms, such as periwinkles and crabs, either burrow in the sand or hide in crevices of rocks.

The distribution of organisms within a rocky intertidal or a sandy shore can vary according to their tolerance to changes in abiotic environmental factors (e.g., temperature and salinity) and in response to biotic factors (e.g., predation). The **zonation** that results can be very striking particularly in the rocky intertidal. Biologists have characterized these zones based on the distribution of common organisms (Figure 13-1).

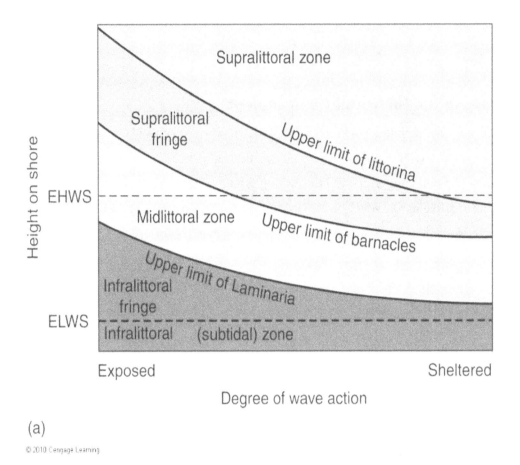

(a)

Figure 13-1 Vertical intertidal zones

The **supralittoral fringe** (upper intertidal or splash zone) is almost always exposed to the air but receives spray from wave action and may be covered during the highest of high tides (EHWS: extreme high water of spring tides). The lower border is marked by the upper limit of most barnacles and the upper border is marked by the upper limit of small snails called periwinkles (*Littorina*). The **midlittoral** (middle intertidal) is regularly exposed at low tide and covered by high tide and is between the supralittoral and the infralittoral fringe. The **infralittoral fringe** (lower intertidal) extends from the area inhabited by the type of brown algae called kelp (*Laminaria*) to the area exposed during the lowest of low tides (ELWS: extreme low water of spring tides). The **infralittoral zone** (subtidal) is never exposed even at low tide (Figure 13-2). Zonation in a soft-bottom intertidal area is not as easy to see immediately, but organisms are distributed based on such factors as the tidal range, sediment type and resulting moisture content of the sediment.

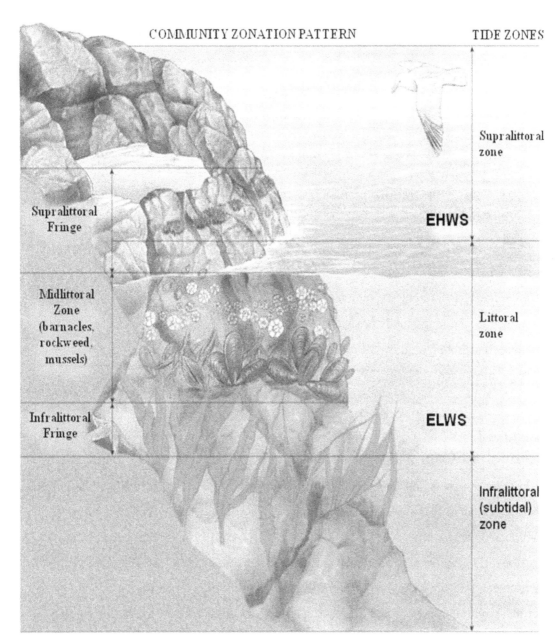

Supralittoral zone

Supralittoral Fringe

EHWS

Midlittoral Zone (barnacles, rockweed, mussels)

Littoral zone

Infralittoral Fringe

ELWS

Infralittoral (subtidal) zone

Figure 13-2 Rocky shore zonation

Materials

- Light meter
- Thermometer
- Refractometer
- GPS (optional)
- 0.5 m² quadrat
- 100 m transect rope or measuring tape
- colorful string
- clipboards
- pencils
- reference charts and identification keys
- shovel (if sampling muddy or sandy shores)
- sieve (if sampling muddy or sandy shores)
- large plastic garbage bags (if sampling muddy or sandy shores)

Exercise 1
Habitat and Community Profile

In order to understand the several adaptations of the organisms that reside in the intertidal zone, we will create a community profile by making a transect. In this community profile we will examine: (1) various physical and chemical conditions, and (2) organisms present.

At low tide, the transect should be perpendicular to the water. Depending on the type of environment, the length may vary. Determine the length so that it is representative of the entire intertidal community. If not sure ask your instructor for advice. Place a meter tape along the transect.

Safety notes:

Never turn your back on the ocean. If you are sampling at a rocky shore be aware that algal communities can make the rocks very slippery. If you sampling in a mud flat, do not proceed into areas where you begin to sink. Be aware of the incoming tide and be vigilant so you do not get cut off. Never climb on large logs at the water line, they can roll and shift suddenly.

Your instructor will provide additional safety notes regarding organisms to be wary of in your area.

Physical and Chemical Characteristics

1. Get into groups of 3 - 4 students. Select and collect data at 10 sites, determined by increments as equal as possible along a transect line. Begin with the Site 1 at the low tide line (if possible) and work your way back to the high tide line.

2. Identify bottom type (rocky, muddy or sandy) at each site and record in Table 13-1.

3. At each site take a reading with the light meter and the thermometer (both air and water if appropriate). If possible (e.g., in a tide pool) use the refractometer to take a measurement of the salinity. Record GPS coordinates (optional). Record data in Table 13-1.

4. Lay your quadrat down at each of your selected sample sites then record the identity and the frequency of the organisms within this plot. If you have a site where one particular organism is extremely numerous it is okay to estimate. For example if your site was full of barnacles of the same species you could divide your quadrat into quarters and only count the barnacles in a quarter of the area-then multiply by 4 to get an estimate for your table. Record data in Table 13-2.

    If you are sampling at a **rocky shore**, make sure you gently move aside any mats of algae and look under the rocks to find the organisms. Make sure you replace any rocks you turn over and return any organism to their original locations.

    If you are sampling at a **sandy shore or a mud flat**, dig down 50 cm (your instructor may alter the depth to suit your field site) in your quadrat. Sieve the sediment onto the plastic sheet placed outside the quadrat and count the organisms. Make sure your replace all the sediment and the organisms. To account for the increased time it takes to sample a soft-bottom intertidal zone, your instructor may adjust the number of sample sites)

5. Proceed to Site 2, assuming that Site 1 was at the low tide line (make appropriate adjustments if it was not) use the transect tape and record the horizontal distance from Site 1. Record in Table 13-1. (When you repeat this procedure for Site 3 etc. you will have the distance between sites, you will need to add them to get the distance from low tide line for your data table).

6. At Site 2 use the transect tape and the colored string to determine your elevation or vertical distance from low tide (Site 1). Start at the site that is at the height of low tide. When continuing to the next site, measure the difference in height. To determine the difference in

height use a string long enough to be stretched between two sites. One team member will then hold the string on the ground at the highest point, and another member will hold the string up until the first member says the string is completely horizontal. Following, a third team member will measure the elevation of the string at the lower site with a meter tape. Record in Table 13-1

(When you proceed from Site 2 to Site 3 etc., you will measure the difference in elevation between the two. Remember to add or subtract appropriately to get the actual vertical distance change from the low tide line)

7. Use the graph paper to draw the elevation pattern found along your transect. Plot the horizontal distance from the low tide line on the x-axis and the vertical distance (elevation) from the low tide level on the y-axis.

8. Use the frequency data collected for organisms found at each site to try and determine the tidal zone (Figure 13-2) for each of your sampling sites.

Table 13-1 Physical characteristics

	Site 1	Site 2	Site 3	Site 4	Site 5	Site 6	Site 7	Site 8	Site 9	Site 10
Substrate type										
distance from Low tide (m) horizontal										
distance from Low tide (m) vertical										
Temperature (°C) air										
Temperature (°C) water										
Light meter reading (lux)										
Salinity (ppt)										
GPS coordinates (optional)										

Table 13-2 Organisms

Identity and Frequency	Site 1	Site 2	Site 3	Site 4	Site 5	Site 6	Site 7	Site 8	Site 9	Site 10

Identity and Frequency	Site 1	Site 2	Site 3	Site 4	Site 5	Site 6	Site 7	Site 8	Site 9	Site 10

Questions

1. Name three stressors that organisms experience living in intertidal zones.

2. What kind of adaptations do organisms use to cope with these stressors?

3. Did you observe differences in the invertebrate communities found at different sites?

4. How did the physical and chemical characteristics changes from your first site to your last site?

5. If the physical and chemical characteristics did change from sampling site to sampling site, are there any additional factors that you need to consider before interpreting your data (e.g., if you started sampling at 9am and finished at noon)?

6. Did you observe vertical zonation?

7. Based on the distribution of organisms which of your sites would you characterize as being in the supralittoral? Midlittoral? infralittoral fringe? Infralittoral zone?

Suggested Reading

Bertness, M. D. 2007. Atlantic Seashores, Natural History and Ecology. Princeton University Press. 431 p.

Ricketts, E. F., Calvin, J. and J. W. Hedgepeth. 1985. Between Pacific Tides. Fifth Edition. Stanford University Press. 652 p.

Glossary of Terms

abiotic: Any part of the environment that can be considered non-living.

active continental margin: Location where the leading edge of a continent collides with an oceanic plate.

biotic: Any part of the environment that can be considered living.

competition: Active demand by two or more organisms for a material or condition, so that both are inhibited by the demand. Interspecific competition would indicate competition between organisms of different species, while intraspecific would indicate the competition between organisms of the same species.

community: Well-defined assemblage of plants and/or animals, clearly distinguishable from other such assemblages.

ecosystem: Community of different species interdependent on each other together with their non-living environment, which is relatively self-contained in terms of energy flow and distinct from neighboring communities.

infralittoral fringe: Region from the area that is exposed at the lowest low tide to the area inhabited by kelp.

infralittoral zone: Subtidal area that is never exposed by low tide.

intertidal zone: Area of a shore that is inundated during high tide and exposed during low tide.

inundation: Being covered by water.

littoral zone: The part of the area between the high- and low-water marks in which photosynthesis/plant life can be sustained.

midlittoral zone: Area regularly exposed by low tide and covered by high tide.

passive continental margin: Location where sea and land meet where no plate subduction or plate collision is taking place.

rocky shoreline: Shoreline where the top layer of material is dominated by rocks.

soft-bottom shoreline: Shoreline where the majority of the top layer is comprised of soft sedimentary materials.

zonation: The separation of living things in a particular habitat into bands or particular areas.

Lab 14
Estuaries

Objectives

Upon completion of this lab exercise, you should be able to:

1) Understand the layering that occurs within estuary ecosystems due to water density differences.

2) Understand the relationship between water temperature, density, and salinity.

3) Understand salinity tolerance ranges and where an organism can be found in an estuary.

Environment of Estuaries

Estuaries are dynamic, semi-enclosed inland areas of water in which freshwater and saltwater combine. Freshwater has relatively low concentrations of **dissolved minerals** and **salts** compared to the water in the oceans. Approximately 99% of the solutes in ocean water are composed of six elements and compounds: chlorine (Cl^-), sodium (Na^+), sulfur (SO_4^{-2}), magnesium (Mg^{2+}), calcium (Ca^{2+}), and potassium (K^+). Trace **ions** and minerals include iron, manganese, cobalt, copper, mercury, and even gold.

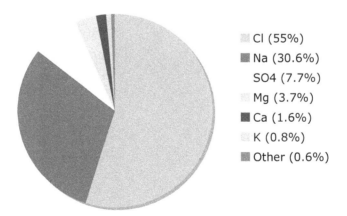

- Cl (55%)
- Na (30.6%)
- SO4 (7.7%)
- Mg (3.7%)
- Ca (1.6%)
- K (0.8%)
- Other (0.6%)

Figure 14-1 Composition of Ocean Water

The composition (salinity) of ocean water causes the **density** to be higher than that of freshwater (Figure 14-1). The density of an object or a liquid is determined by dividing the mass of the object by its volume. The density of pure water is 1 g/cm^3. Ocean water has a higher density, 1.02-1.03 g/cm^3, and thus weighs 2-3% more than freshwater. Therefore, when ocean water and freshwater come together, the denser ocean water tends to "sink" under the freshwater.

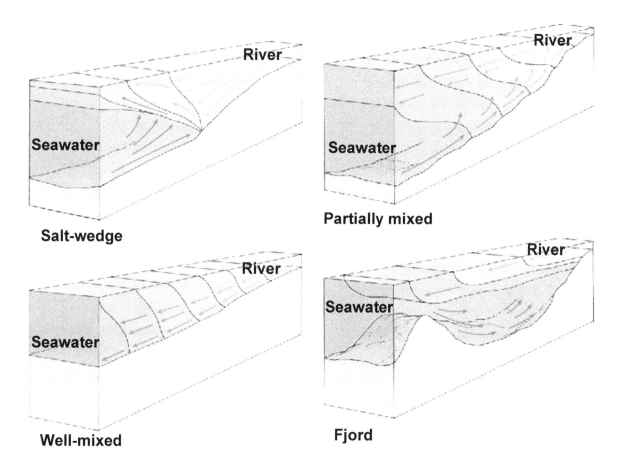

Figure 14-2 The four major types of mixing patterns in estuaries.

Estuaries are areas where rivers empty into the ocean; and because of the action of tides, ocean water sometimes travels upstream. This constant mixing creates strong, complex salt and temperature **gradients** along the length of the estuary. There are four main categories of estuaries, based on general water mixing patterns: **salt wedge**, **partially mixed**, **well-mixed**, and **fjord** (Figure 14-2). The freshwater and saltwater are **stratified** in these systems by energy derived from tidal action, bottom profile, geomorphology, wind, and water runoff. These factors interact, mixing the waters and producing different salinity and temperatures profiles. These ecosystems have the ability to change quickly, placing both physiological and chemical demands on the organisms present. Typically, there are several options for animals when they encounter harsh or stressful environmental conditions: 1) migrate to an area with more favorable conditions; 2) take up residence and adapt or resist local conditions; 3) if the stress is too great, the animal will die.

While estuaries can be a very challenging habitat they are also very productive. Estuaries are nurseries for a large number of species. Relatively few species have adaptations to flourish in estuaries in all life stages, often the juveniles have different adaptations than those of adults.

In this lab you will investigate the relationship between salinity, temperature and density. You will also use the data sets given to explore the different kinds of estuaries and the possible distribution of organisms within an estuary.

Materials

- Small aquarium (~10L) with grooves to receive a divider—longer and shorter aquariums work best
- Plexiglass divider cut to fit aquarium width
- Two 4.5-5 L containers for preparing water for aquarium
- Refractometer—can substitute by hydrometer as an alternative
- Thermometer (°C)
- Sodium chloride (NaCl)
- Food coloring (blue, green, red and yellow)
- Water (tap water is fine)
- Four 250-mL graduated cylinders
- Sodium chloride (NaCl)
- Top loading balance (g)
- Three 9-cm Pasteur pipettes
- Calculator

Exercise 1
Salinity

1) Get into groups of 3-4 students. Gather materials including one aquarium per group.

2) Insert the Plexiglass in the middle of the aquarium, making sure it fits snuggly against the sides and bottom. The purpose is to form a barrier that will not allow the water on each side of the aquarium to mix readily.

3) Using the graduated cylinders measure 4 L of cool tap water (between 10 and 17 °C) into your first 4.5-5 L container.

4) Using the balance, weigh out 100 g of NaCl into the large weigh boat and mix 100g of NaCl into your first 4.5-5L and mix until all the salt is completely dissolved.

5) Using the graduated cylinders measure 4 L of cool tap water (between 10 and 17 °C) into your second 4.5-5 L container.

6) Add 10–12 drops of blue food coloring to the first container of salt water and mix well. Add 10-12 drops of yellow food coloring to the second container of freshwater and mix well.

7) Record the temperature of each solution in °C. Make sure the temperature of each solution is approximately the same. Adjust if necessary. Record data in Table 14-1.

Table 14-1 Density, Temperature, and Salinity Data

Water Sample	Salinity ppt	Temperature °C	Density (from Table 14-2)
Blue salt water			
Yellow tap water			
Observations			
Green cold water			
Red warm water			
Observations			

8) Using the refractometer, measure the salinity of both solutions. Record data in Table 14-1.

9) Salinity is measured in ppt. Use Table 14-2 to convert the data for salinity and temperature in °C into density. First identify the column that corresponds to the temperature of the water; then find the salinity reading from the refractometer. Once you find the salinity under the proper temperature, move to the left column, which will be the density of your water solution. Record data in Table 14-1. Simultaneously have one member of the group slowly pour the blue salt water in one end of the tank and another member of the group slowly pour the yellow tap water in to the other end of the tank. Wait about 10-15 seconds until the water is relatively still.

Table 14-2 Relationship between water temperature, density, and salinity.

Density (g/cm³)	Temperature (°C)							
	10	11	12	13	14	15	16	17
1.012	14.7	14.8	14.9	15	15.2	15.4	15.7	15.8
1.013	15.8	16	16.2	16.3	16.5	16.7	17	17.1
1.014	17.1	17.3	17.5	17.7	17.8	18	18.3	18.6
1.015	18.4	18.6	18.8	19	19.1	19.3	19.6	19.9
1.016	19.7	19.9	20.1	20.3	20.4	20.6	20.9	21.2
1.017	21	21.2	21.3	21.6	21.7	22	22.2	22.5
1.018	22.3	22.5	22.6	22.9	23	23.3	23.5	23.8
1.019	23.6	23.8	23.9	24.2	24.3	24.6	24.8	25.1
1.020	24.8	25.1	25.2	25.5	25.6	25.9	26.1	26.4
1.021	26.1	26.4	26.5	26.8	26.9	27.2	27.4	27.7
1.022	27.4	27.7	27.8	28.1	28.2	28.5	28.7	29
1.023	28.7	28.9	29.1	29.4	29.5	29.8	30	30.3
1.024	30	30.2	30.4	30.6	30.8	31.1	31.3	31.6
1.025	31.3	31.5	31.7	31.9	32.1	32.4	32.6	32.9

10) Lift the Plexiglass 5 cm at a time, very slowly, and observe how the solutions mix. Determine which one moves in further, which one is on the bottom. Note layering. Do you see an area of mixing? How do you know?

11) Dispose of the contents of your aquarium according to the directions given by your instructor Repeat steps 2-11 with 4L of cold water (~ 5°C) and warm water (~ 50°C). Add the green food coloring to the cold water and red food coloring to the warm water. Record data in Table 14-1.

Questions

1. Based on your observations, which is denser, the salty or the freshwater? The cold or the warm water? Why?

2. What happened with the two volumes of water in your aquarium?

3. Where would you expect to find the saltier water in an estuary?

Exercise 2
Effects of Temperature and Salinity on Stratification

In the previous exercise you observed that both temperature and salinity affect water density. In the marine environment both the salinity and the temperature can vary vertically in the water column. This variation can result in layering or stratification. Depending on the season and the region, the top layer of the water is often warmer, the bottom layer is colder and then there is a region of rapid temperature change called a **thermocline**. Similarly the salinity can vary forming a **halocline**. The combination of the effect of temperature and salinity on the density of water can result in layering and a zone where the density changes rapidly called the **pycnocline**. To investigate the phenomena four different densities of solution will be made to represent stratifications within an estuary ecosystem. The solutions will be labeled A–D and dyed differently with food coloring for observation.

1. Get into groups of 3-4 students. Gather four 250 mL graduated cylinders and label them A-D.

2. Using a balance, weigh out 50 g of NaCl and place in the bottom of the graduated cylinder. Add tap water until the volume in the cylinder is 200 mL. Mix well. Label this cylinder A.

3. Remove 125 mL of the solution from cylinder A and put in a second graduated cylinder. Add tap water until the volume in this cylinder is 200 mL. Make sure the temperature remains approximately the same as in the original solution. Mix well. Label this cylinder B.

4. Remove 125 mL of the solution from cylinder B and put in a third graduated cylinder. Add tap water until the volume in this cylinder is 200 mL. Make sure the temperature remains approximately the same as in the original solution. Mix well. Label this cylinder C.

5. Add warm tap water (make sure it is 10-15 °C warmer than your original solutions) to the last cylinder until the volume in this cylinder is 200 mL. Label this cylinder D.

6. Do not add any food coloring to cylinder A, add 7 drops of yellow food coloring to cylinder B, add 7 drops of blue food coloring to cylinder C and 7 drops of red food coloring to cylinder D

7. Once all the solutions have been prepared, fill a Pasteur pipette with solution B. Do the same with solutions C and D. Introduce solution B into solution A by **slowly** streaming it down the sides of graduated cylinder. A. Add approximately 40 ml of solution B. Repeat in the following sequence:

   - C on top of B
   - D on top of C

As you add the solutions, record your observations.

Questions

1. How well did the solutions mix (if they mixed at all)?

2. Were there any two that mixed better than the others?

3. Why did some of the solutions remain stratified?

4. What would happen if you stirred the solutions?

5. What natural processes could stir the waters in an estuary?

6. What human activities stir the waters in an estuary?

Exercise 3
Animal Adaptations to Salinity Stratification

Work these problems on your own and then, as a group, discuss results.

Density = Mass/Volume (remember that cm^3 is equivalent to ml)

Problem 1. Marine biologists collected water samples at various depths at two sites in an estuary.

Calculate the densities of the samples and determine the type of estuary using Figure 14-2.

Site 1			
Depth (m)	Mass (g)	Volume (ml)	Density
0	30.31	30	
1	25.15	25	
2	45.27	45	
3	30.19	30	
Site 2			
0	40.60	40	
1	25.39	25	
2	30.48	30	
3	15.24	15	

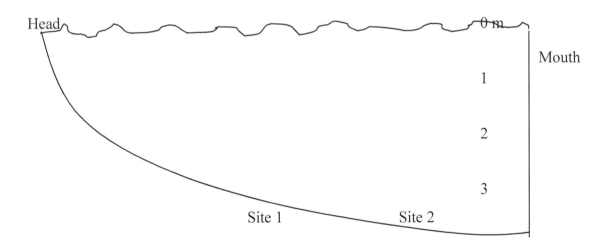

Problem 2. The biologists moved a mile south and collected water samples at various depths at two sites in the next estuary. Calculate the densities of the samples and determine the type of estuary using Figure 14-2.

Site 1			
Depth (m)	Mass (g)	Volume (ml)	Density
0	18.02	18	
1	52.10	52	
2	31.12	31	
3	28.17	28	
Site 2			
0	24.05	24	
1	15.06	15	
2	44.66	44	
3	37.81	37	

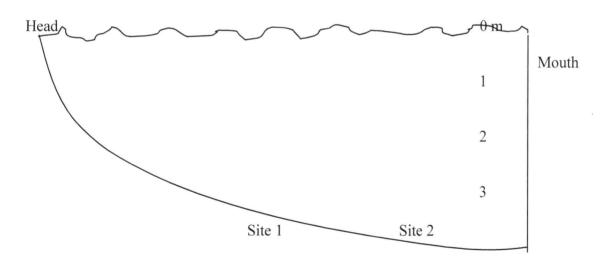

Problem 3. The marine biologists at a third estuary also collected water temperature at the various depths. Determine the salinity using Table 14-2. Using the range information for different species in Table 14-3 where would you expect to find the various species in this estuary? Remember that organisms often have different salinity preferences at different life stages and that temperature can be an important factor in determining tolerance.

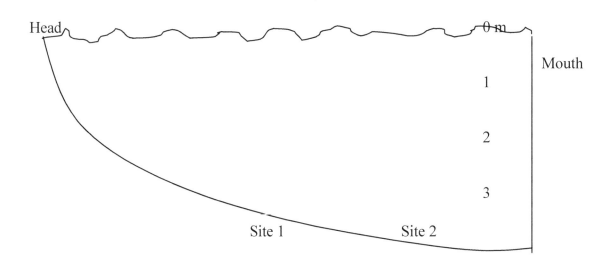

Site 1			
Depth (m)	Density	Temperature °C	Salinity
0	1.015	17°	
1	1.016	16°	
2	1.020	11°	
3	1.021	11°	
Site 2			
0	1.020	17°	
1	1.022	14°	
2	1.024	11°	
3	1.025	10°	

Table 14-3 Salinity preference range (from Patillo et al. 1995)

Species	Salinity Range	Depth	Site
Bay Squid (*Lolliguncula brevis*)	20-36 ppt		
Blue crab (*Callinectes sapidus*)	23 – 33 ppt		
American Oyster (*Crassotrea virginica*)	2 – 40 ppt		
Gulf Menhaden (*Bevorrtia patronus*)	25 – 37 ppt		
Gulf Killifish (*Fundulus grandis*)	0 – 45 ppt		
Pink Shrimp (*Penaeus duorarum*)	27-35 ppt		
Code Goby (*Gobiosoma robustum*)	10-30 ppt		
Atlantic Rangia (*Rangia cuneata*)	1-15 ppt		

Questions

1. In problem 1 what type of estuary is pictured?

2. In problem 2 what type of estuary is pictured?

3. Based on the information given, would you be confident that you could predict where you would find the organisms listed in the estuary pictured?

Suggested Reading

Dyer, K. R. 1998. *Estuaries: A Physical Introduction.* Second Ed., John Wiley & Sons, New York.

Kennish, M. J. 2003. *Estuarine Research, Monitoring, and Resource protection.* CRC Press, Boca Raton, Florida.

Pattillo, M., Rozas, L. P., and R. J. Zimmerman. 1995. A review of the salinity requirements for selected invertebrates and fishes of U.S. Gulf of Mexico estuaries. National Marine Fisheries Service, Southeast Fisheries Science Center, Galveston Laboratory. A Final Report to the Environmental Protection Agency. 62 pp.

Glossary of Terms

density: The mass of a substance in a given volume of that substance.

estuaries: A region where freshwater is mixed with saltwater.

fjord: A deep valley cut into the coastline by glaciers and filled with a mixture of freshwater and saltwater.

halocline: a layer in ocean water where the salinity changes rapidly

ions: An atom or a molecule that carries an electrical charge.

partially mixed: An estuary that has a strong surface flow of freshwater and a strong influx of seawater.

pyncocline: a boundary layer in ocean water where the density changes rapidly.

salinity: A measure of the concentration of dissolved inorganic salts in water.

salt: A compound composed of two ions, usually remains dissolved in a solution after an acid/base reaction has occurred e.g., NaCl

salt wedge: The angled boundary between saltwater and freshwater in an estuary that occurs when the rapid flow of river water prevents the saltwater from mixing with the freshwater.

thermocline: a layer in ocean water where the temperature changes rapidly

well-mixed: An estuary in which river flow is low and tidal currents play a major role in the circulation of the water, resulting in a seaward flow of water and a uniform salinity at all depths.

Lab 15

Fishing Down the Food Chain

Objectives

Upon completion of this exercise, you should be able to:

1) Describe the factors that contributed to increases in landings of targeted species over the past 70-80 years.

2) Describe the difference between a **target species** and **bycatch**.

3) Describe how overfishing of one targeted species leads to changes in fishing practices that increase pressure on other species.

4) Explain how the harvesting of species at lower trophic levels affects the repopulation of commercially hunted species at higher trophic levels.

Background: Fisheries

A recent **FAO** report indicated that the global fishing fleet consisted of over 4.3 million vessels (2010). Fisheries, both capture and aquaculture, are significant sources of protein for much of the world's population. For approximately 1.5 billion people, fish and seafood consumption is responsible for at least 20% of their animal protein intake and in some small island nations the figure is closer to 90%. Fisheries are also an important economic engine for many countries, it has been estimated that directly or indirectly 8% (or about 540 million people) of the world's population is dependent on some facet of the fisheries for their livelihood (FAO 2010). Unfortunately while the world's population continues to grow, the harvest from the ocean has peaked (Figure 15-1). Fisheries around the world are seeing a decline in individual stocks. The collapse of some fishing stocks has been dramatic and been widely reported in the press (e.g., the closure of the Atlantic Cod fishery) but others have received little notice.

Changes in technology, increases in fishing effort and a broadening of the number of species taken commercially have masked the decline of many individual fish stocks. Fishers from Europe have been traveling as far as the Grand Banks since the 1500's but fishing effort was in part limited by vessel capacity and storage limitations. In the 1950's the first factory freezer trawler was produced and this began the era of large fleets of factory ships that could travel far from home and process a huge catch. In response to domestic concerns about the fishing pressure

190   Fishing Down the Food Chain

of foreign fleets, the United Nations *Convention on the Law of the Sea* resulted in the establishment of **exclusive economic zones (EEZ)**. This led to greater investment in domestic fishing fleets without respect for what kind of take could be sustained by local fish stocks. Additionally, changes in technology such as sonar and spotter planes, allowed fishing boats to hunt very efficiently. The increased use of bottom trawlers that destroy benthic habitats and leave a 'mud trail' visible from satellites and change the optical properties of the water column, and the use of passive gear in the open ocean (e.g., drift nets) culminated in ecological disaster for species like the Atlantic Cod. **Bycatch**, the sea life captured along with the higher value fish or seafood that the boat will keep for sale, added to these pressures. For some fisheries the bycatch to catch ratio is very high (over 10:1) so increased fishing pressure means even greater increases in the pressure on organisms unlucky enough to be near the targeted species. Long lines and drift nets (now banned by the United Nations) are responsible for the unintended deaths many dolphins, sea birds and sea turtles.

As some fisheries have declined or closed, pressure has shifted to other stocks. The increased tonnage in total landings after the collapse of some major fisheries is often due to greater targeting of invertebrates, small mid-pelagic fish and species that would have been considered 'trash' previously. The Alaskan Pollock is a close relative of the Atlantic Cod but has a much stronger flavor. The Alaskan Pollock fishery was not fished primarily by American vessels until the late 1980's but now it makes up a significant portion of the Gulf of Alaska fisheries catch. World wide it is one of the top 10 (FAO 2010) species harvested by volume. Another response to the decline of certain species like salmon, was the rise of aquaculture. However, fish farming for salmon or trout or tuna requires that these predatory organisms be fed high protein diets. This is generally fish pellets, made out of processed captured smaller fish. It takes many pounds of fish pellets to produce one pound of commercially farmed carnivorous fish. The farming of carnivorous fish lead to a market for captured species that could be turned into fish pellets adding another stress to the marine food web rather than relieving it.

Fishing down a food web can have ripple effects in marine ecosystems. Some species, like the mid-pelagic fish, are responsible for important energy transfers between phytoplankton (producers) and seabirds and marine mammals (secondary/tertiary consumers). Some food chains have been fished so intensively that they have lost much of their complexity, which has left the organisms to survive in a more linear system that is more vulnerable to other stressors, such as temperature fluctuations in the environment and disease. In some regions the food web has shifted significantly due to overfishing. The exploitation of anchovy and sardine stocks off the coast of Namibia reduced the competition for food and now the jellyfish dominate (Lynam et al.

2006). The efficient selective removal of large fish often alters the age structure of a population significantly. In many species of rockfish for example it is known that it is the large females that produce the most eggs. Not only do they produce the most eggs but their offspring are more likely to survive-so a population with an altered age structure e.g., one in which the most reproductively viable individuals have been removed and served for dinner, is going to be less resilient and have a harder time recovering even if the fishing pressure is reduced. The conventional wisdom has been that it was the species at the top of the food web that were the most vulnerable to collapse. However recent analysis suggests that this is not the case. Fish lower on the trophic scale, but that live in groups that are easy to catch in large numbers, are also extremely vulnerable (Pinsky et al. 2011). Other researchers suggest that a shift in some systems to being dominated by a large biomass of microbes, rather than fish and invertebrates, leaves the system more vulnerable to pollution and disease (Jackson et al. 2001).

Conservation efforts are increasing shifting towards an ecosystem approach rather than a species approach. The lobby for the establishment of marine protected areas and marine reserves has had some significant successes, and a properly designed series of marine protected areas or reserves can increase the availability of commercial species outside of the zone in which conservation takes precedence, by serving as protected nurseries for members of many species which then move out to surrounding areas. Better management practices of areas being fished, as well as areas of conservation, can help prevent degradation of marine food webs. In some fisheries there has been substantial success via establishing individual quotas that fishers are able to catch at the point in the season when the supply and demand relationship is most favorable to them, rather than an overall catch limit for a fleet that promotes a rapid scramble competition at the beginning of the fishing season. Education of consumers, by programs like the Monterey Bay Aquarium Seafood Watch, can be extremely valuable by helping to move market forces, with an increasing demand for sustainably harvested species and ones not in rapid decline, rather than species that are overexploited and near collapse. As dire a story as the saga of fishery collapse may be, it is important to celebrate success stories, like environmental group and consumer pressure leading to Dolphin safe tuna captured in a way that does not kill excessive numbers of cetaceans, and the turtle excluder devices (TEDs) which allow sea turtles a way out of shrimp nets before they drown, which therefore have dramatically reduced the number of sea turtles killed in the Gulf shrimp industry and elsewhere.

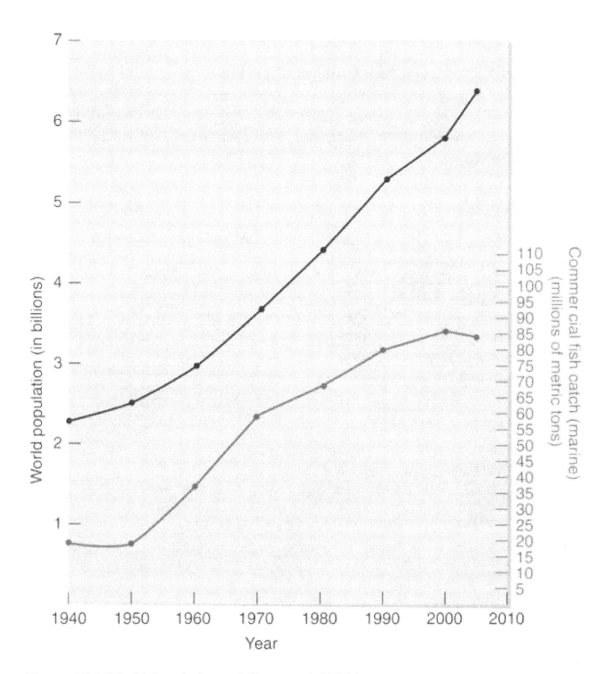

Figure 15-1 World Population and Commercial Fishing

Exercise 1

Analysis of Commercial Fish Catches Over Time

1. Using the data provided in Table 15-1 plot the number of metric tons of landings in the United States from 1950 to 2010. Put the year on the x-axis and the number of metric tons caught on the y-axis.

Table 15-1 Commercial Landings US Fisheries (Personal communication from the National Marine Fisheries Service, Fisheries Statistics Division, Silver Spring, MD)

Year	Metric Tons	Year	Metric Tons	Year	Metric Tons
1950	2,258,042.80	1975	2,303,027.30	1995	4,487,444.20
1955	2,238,972.90	1980	2,974,843.00	2000	4,147,172.40
1960	2,304,603.90	1985	2,894,262.30	2005	4,405,514.10
1965	2,214,960.30	1990	4,457,084.10	2010	3,741,554.10
1970	2,239,812.60				

a) What is the trend over time?

b) When did landings peak in the United States?

c) How does this compare to the world commercial marine catch shown in Figure 15-1 (note the units for the World data are millions of metric tons compared to metric tons for the US data)?

2. What is happening in different fisheries over time? Using the data provided in Table 15-2 plot the number of metric tons of landings of Atlantic Cod (*Gadus morhua*) in all states from 1970 to 2010. Put the year on the x-axis and the number of metric tons on the y-axis.

Table 15-2 Commercial Landings Atlantic Cod (Personal communication from the National Marine Fisheries Service, Fisheries Statistics Division, Silver Spring, MD)

Year	Metric Tons	Year	Metric Tons	Year	Metric Tons
1950	26,077.50	1975	25,462.10	1995	13,632.70
1955	16,139.80	1980	53,422.20	2000	11,371.70
1960	18,317.10	1985	37,432.00	2005	6,311.00
1965	16,350.90	1990	43,624.40	2010	8,039.40
1970	24,143.20				

a) What is the trend over time?

b) When did the Atlantic Cod landings peak in the United States?

c) How do the peaks compare between the overall US fisheries data and the Atlantic Cod?

d) If they are different, why do think this is the case?

3. Is there evidence that fishing pressure on different species has changed over time? Using the data provided in Table 15-3 plot the number of metric tons of landings of Alaskan Pollock (*Theragra chalcogramma*) (also known as Walleye Pollock, this is now used to make **surimi**). Put the year on the x-axis and the number of metric tons on the y-axis.

Table 15-3 Commercial Landings Alaskan Pollock (Personal communication from the National Marine Fisheries Service, Fisheries Statistics Division, Silver Spring, MD)

Year	Metric Tons	Year	Metric Tons	Year	Metric Tons
1950	*	1975	20.8	1995	1,259,776.00
1955	*	1980	1,445.30	2000	1,182,436.90
1960	*	1985	46,143.20	2005	1,547,358.80
1965	66.4	1990	1,408,670.90	2010	883,416.30
1970	193.6				

*no records of commercial landings

a) What is the trend over time?

b) When did the Alaskan Pollock landings peak in the United States?

c) How do the peaks compare between the Atlantic Cod and the Alaskan Pollock?

d) What are some possible reasons for the lack of data for 1950, 1955 and 1960?

4. Look at the data provided in Tables 15-4 and Table 15-5, giving landing by species for Oregon in 1950 and 2010.

Table 15-4 Commercial Landings for Oregon 1950 (Personal communication from the National Marine Fisheries Service, Fisheries Statistics Division, Silver Spring, MD)

AFS species name	Metric tons	AFS species name	Metric tons
BASS, STRIPED	16.6	SALMON, CHINOOK	3843.0
CLAM, PACIFIC RAZOR	58.2	SALMON, CHUM	256.8
COD, PACIFIC	19.5	SALMON, COHO	1234.5
CRAB, DUNGENESS	2842.6	SALMON, SOCKEYE	56.7
CRAYFISHES	14.7	SHAD, AMERICAN	593.0
FLATFISH	165.4	SHARK, DOGFISH	318.9
HAKE, PACIFIC	0.63	SHARK, SOUPFIN	74.7
HALIBUT, PACIFIC	322.3	SHARKS	21.5
HERRING, PACIFIC	23.77	SKATES	7.8
LINGCOD	399.7	SMELT, EULACHON	319.2
OCTOPUS	0.04	SOLES	5586.7
OYSTER, PACIFIC	442.7	STURGEONS	82.7
ROCKFISHES	2596.3	SURFPERCHES	47.1
SABLEFISH	202.6	TROUT, RAINBOW	419.8

a) How many landing categories (i.e., species) are listed in 1950?

b) How many invertebrate species are listed in 1950?

Table 15-5 Commercial Landings for Oregon 2010 (Personal communication from the National Marine Fisheries Service, Fisheries Statistics Division, Silver Spring, MD)

AFS species name	Metric tons	AFS species name	Metric tons
ANCHOVY, NORTHERN	130.9	ROCKFISH,YELLOWEYE	0.7
CABEZON	23.5	ROCKFISH,YELLOWTAIL	90.4
CLAM, BUTTER	1.2	ROCKFISHES	854.8
CLAM, PACIFIC RAZOR	9.9	SABLEFISH	2843.7
CLAM, PACIFIC, GAPER	1.3	SALMON, CHINOOK	976.4
COCKLE, NUTTALL	44.6	SALMON, CHUM	0.6
COD, PACIFIC	54.8	SALMON, COHO	265.4
CRAB, DUNGENESS	7170.7	SALMON, SOCKEYE	0.5
CRAB, RED ROCK	1.1	SARDINE, PACIFIC	20295.0
CRABS	2.4	SEA URCHINS	113.8
ECHINODERM	0.1	SHAD, AMERICAN	4.1
FINFISHES, GENERAL	32.3	SHARK, BLUE	0.06
FLOUNDER ARROWTOOTH	2209.3	SHARK SPINY DOGFISH	58.9
FLOUNDER, STARRY	10.0	SHELLFISH	0.02
FLOUNDER,PACIFIC	90.3	SHRIMPBLUE MUD	1.06
GREENLING, KELP	18.3	SHRIMP, GHOST	30.7
GRENADIERS	41.5	SHRIMP, OCEAN	14256.0
HAGFISHES	786.7	SHRIMP, PENAEID	7.8
HAKE, PACIFIC	25862.3	SKATES	918.1
HALIBUT, PACIFIC	84.4	SMELT, EULACHON	1.3
LINGCOD	77.1	SOLE, BUTTER	0.05
MACKEREL, CHUB	11.5	SOLE, CURLFIN	0.1
OCTOPUS	2.1	SOLE, DOVER	6726.2
OYSTER, PACIFIC	188.1	SOLE, ENGLISH	124.1
ROCKFISH, BLACK	101.3	SOLE, FLATHEAD	1.5
ROCKFISH, BL-AND-YEL	0.01	SOLE, PETRALE	504.0
ROCKFISH, BLUE	3.3	SOLE, REX	355.2
ROCKFISH, BROWN	0.03	SOLE, ROCK	0.7
ROCKFISH, CANARY	3.6	SOLE, SAND	60.0
ROCKFISH, CHINA	4.8	SQUIDS	42.4
ROCKFISH, COPPER	0.7	STURGEON,WHITE	60.4
ROCKFISH DARKBLOTCHED	148.0	THORNYHEAD	760.0
ROCKFISH, GOPHER	0.1	TROUT, RAINBOW	11.4
ROCKFISH, GRASS	0.2	TUNA, ALBACORE	4854.8
ROCKFISH,PAC. PERCH	57.5	WOLF-EEL	0.05
ROCKFISH, VERMILION	1.6	YELLOWTAILJACK	0.07
ROCKFISH, WIDOW	29.2		

a) How many landing categories (i.e., species) are listed in 2010?

b) Would changes in fishing pressure be the only reason the number of categories would increase? What other factors do you need to consider when interpreting these data?

c) How many invertebrate species are listed in 2010?

d) Do you see evidence of new fisheries developing?

e) Do you see evidence of organisms in lower trophic levels being targeted?

5. You will now identify a research question.

Possibilities:

- Do the trophic level patterns in the Oregon data match those for other fisheries?

- Did the introduction of a particular kind of gear alter the landings for a particular target species

- What was the effect of the 2010 Deep Water Horizon Oil spill on the 2010 shrimp catch in the Gulf?

- Think of your own!

6. Based on your question develop a hypothesis and use NOAA Marine Fisheries data (the source of all the earlier data used in this lab) to try to test your hypothesis. Choose the closest major fishery to your location or one indicated by your instructor. To access the fisheries data go to the NOAA Fisheries Office of Science and Technology website at http://www.st.nmfs.noaa.gov.

    a. Scroll down to the flow chart and click on the lower left item, labeled **Fisheries Statistics**.

    b. Look for the item labeled **Commercial Fisheries** and click on it.

    c. Click on the **Annual Landings** item shown in the menu. This will take you to a specific site where you can find all commercial fish landing statistics that have been archived by the National Marine Fisheries.
       You can also choose **Landings by Gear** or **Annual Landings with Group Subtotals**.

    d. There will be options for **selecting species, year range** and **geographical area**.
       (to import data directly into an Excel spreadsheet, copy data then paste special as text)

Research Question:

Hypothesis:

Method summary (what data will you need to pull from the website to test your hypothesis?)

    e. Use the website to collect the data need to test your hypothesis.

    f. Graph your results.

    g. Did your data support or allow you to reject your hypothesis?

    h. What conclusions were you able to make about your research question?

Exercise 2
Sustainable Choices

There are many tools available to help the consumer choose seafood that is sustainable e.g., Marine Stewardship council certification (www.msc.org) or the Monterey Bay Aquarium SeaFood Watch program (http://www.montereybayaquarium.org/cr/seafoodwatch.aspx/) that has regional lists of good versus poor seafood choices available. NOAA also has FishWatch (www.fishwatch.gov) to provide information about sustainable choices. The University of Rhode Island (URI) sustainable seafood initiative (http://seagrant.gso.uri.edu/sustainable_seafood/index.html) also has a database and links to others to make good consumer choices. Often a fishery is assessed based on the known population levels of the stock, if the gear used to catch the target species harms the habitat and the levels of bycatch. Other important questions to ask include whether the seafood is wild or farmed and is it local? How it is farmed can be very important because some farming operations are sustainable and some are not. In this exercise you will assess the choices available to the consumer in your community.

1. Get into a group of 3-4 students. Make a list of the possible places to purchase seafood in your community. Include fish markets, independent grocery stores, major chain grocery stores and big box stores (e.g., Wal-Mart, Target). If possible assign one location to each member of the group, try to represent the types of stores as broadly as possible e.g., one student is assigned a chain grocery store, another student is assigned a fish market. (If transportation is an issue some large chains now allow shopping online). Group members will share data.

2. Go to assigned store and find all the seafood products sold (canned, frozen, dried and fresh). Collect data to answer the following questions:

   a. How many different types of seafood were sold?

   b. What were the countries of origin?

   c. Was it possible to assess whether or not the item was a sustainable choice or not? (describe what criteria you used to assess 'sustainability').

   d. Was it possible to purchase seafood from the port that is nearest to your location? i.e., can you buy local?

e. Were there differences between the different kinds of stores (use the group data to answer this question).

Questions

1. You are having a casual conversation about overfishing and mention that if trends continue the way they are now we might all end up eating jellyfish. Obviously, you are using this to make a point. Assuming the other people you are conversing with are not well acquainted with the history of fisheries conservation, write a brief description of the point you were trying to make using the topic of "Fishing Down the Food Chain" as the focus in your explanation.

2. Research five types of technology developed in the past 125 years that have enabled us to get to the point of fishing the oceans out of most of its commercially valuable species. Given that this has been at such a rate that we have to resort to fishing at lower trophic levels in order to still harvest protein from the sea, discuss how each type of technology contributed to the over-harvesting problem.

3. Contrast the differences between the species approach and ecosystem approach to wildlife management. Provide specific examples of each and discuss the pros and cons of the two types of management approaches.

Suggested Readings

FAO. The State of World Fisheries and Aquaculture 2010. Rome, FAO. 2010. 197p.

Jackson, J. M. X. Kirby, W. H. Berger, K. A. Bjorndal et al. 2001. Historical ovefishing and the recent collapse of coastal ecosystems. Science. 293:629-638.

Lynam, C. P., M. J. Gibbon, B. E. Axelsen, C. A. J. Sparks, J. Coetzee, B. G. Heywood, and A. S. Brierly. 2006. Jelly-fish overtake fish in a heavily fished ecosystem. Current Biology 16: 492-293.

NOAA. 2012. FishWatch U. S. Seafood facts. http://www.fishwatch.gov/

Marine Stewardship Council. 2012. www.msc.org

Monterey Bay Aquarium Seafood Watch Program. 2012. http://www.montereybayaquarium.org/cr/seafoodwatch.aspx/

NOAA Fisheries Service. 2012 http://www.nmfs.noaa.gov/

NOAA Fisheries Service. 2012. Office of Science and Technology http://www.st.nmfs.noaa.gov

Northwest Center for Sustainable Resources. 2009. NCSR Marine fisheries series. www.ncsr.org

Pauly, D. and R. Watson. 2009. Spatial dynamics of marine fisheries, p. 501-509 in The Princeton Guide to Ecology, S. Levin (ed). Princeton University Press, Princeton, N.J.

Pauly, D., V. Christensen, J. Dalsgaard, R. Froese and F. Torres. 1998. Fishing down marine food webs. Science. 279: 860-863.

Pinsky, M. L., O. P. Jensen, D. Ricard and S. R. Palumbi. 2011. Unexpected patterns of fisheries collapse in the world's oceans PNAS 108(20): 8317–8322

Roberts, C. 2007. The unnatural history of the sea. Island Press Washington, D. C. 435 p.

Glossary of Terms

bycatch: The noncommercial animals killed during fishing for commercial species; also known as *incidental catch*.

drift nets: A large net composed of sections called *tans* that may stretch as far as 60 kilometers. Drift nets entangle fish, squid, and other marine animals that swim into them.

ecosystem approach to wildlife management: An approach that looks at all components of an ecosystem as interrelated, and as things that must be considered when protecting and restoring natural balances.

Exclusive Economic Zone (E.E.Z.): Area of ocean exclusively controlled by a coastal nation.

fishing effort: A measure of the number of boats fishing, the number of fishers working, and the number of hours they spend fishing.

FAO: Food and Agriculture Organization, a United Nations entity

incidental catch: Noncommercial animals that are killed each year during fishing for commercial species; also known as *bycatch*.

landing: The catch made by a fishing vessel.

species approach to wildlife management: An approach that focuses primarily on the preservation of one species.

stock: Separate population of commercial fishes or shellfishes within a species' geographic range that is assumed to be reproductively isolated from other stocks.

surimi: A product made from the flesh of the Alaskan Pollock (*Theragra chalcogramma*). It is flavored to produce artificial crab, shrimp, and lobster.

sustainable yield: The number of fishes and shellfishes that can be caught over several years without stressing the population.

trawling: A large net dragged along the bottom or in midwater, depending on the catch, by vessels called *trawlers*.

Lab 16

Dead Zones

Objectives

Upon completion of this exercise, you should be able to:

1) Define **dead zone, hypoxia, anoxia,** and **eutrophication**.

2) Discuss the factors that contribute to the development of a dead zone.

3) Discuss some of the effects of dead zones on marine ecosystems.

4) Discuss possible solutions to relevant dead zone water quality issues primarily influenced by human activity.

Hypoxic Zones

Dead Zones can result in areas where the water is so low in dissolved oxygen that many organisms leave or are killed. Along many coastlines all around the world, scientists are finding an increase in areas of **hypoxia** (low oxygen) or **anoxia** (no oxygen). The 2010 Scientific Assessment of Hypoxia in U.S. coastal waters (Committee on Environment and Natural Resources) found that 307 of the 647 coastal and estuarine systems assessed had problems with hypoxia. Hypoxic zones can form due to natural ocean processes or as a result of human activity. Changes in upwelling patterns due to climate change have resulted in areas of seasonal hypoxia off the coast of Oregon. Many other areas of hypoxia are due to **eutrophication**. Eutrophication is initiated when an increase in nutrients that limit growth occurs. This leads to increased growth in algal populations (a bloom). Most aquatic systems are limited in their productivity by the available nitrogen or phosphorus in the water, so increasing the concentration of those nutrients is like using 'fertilizer'. The resulting algal bloom can affect the ecosystem in many ways. It may be unsightly or result in the release of toxins that affect shellfish and other organisms (a harmful algal bloom, **HAB**). The increased growth in the photic zone can affect the amount of light reaching submerged grasses or macroalgae. If the explosion of growth uses up all the nutrients available, or after their short life spans, the algae die. The process of decomposition of dead algae by aerobic bacteria depletes the oxygen in the water and the water becomes hypoxic or even anoxic.

The source of nutrients in the water can vary depending on the watershed. Many areas are naturally high in nutrients but some suffer from nutrient pollution and then **cultural eutrophication** can occur. Historically, **point sources** like waste-water treatment plants were the most important sources of excess phosphorus or nitrogen. In some places this is still the case, but in the United States non-point sources are now more significant. **Nonpoint** sources can include run-off from agricultural operations that use fertilizer, waste products from animal operations, the effect of nitrogen-fixing from leguminous crops, and atmospheric deposition from the burning of fossil fuels

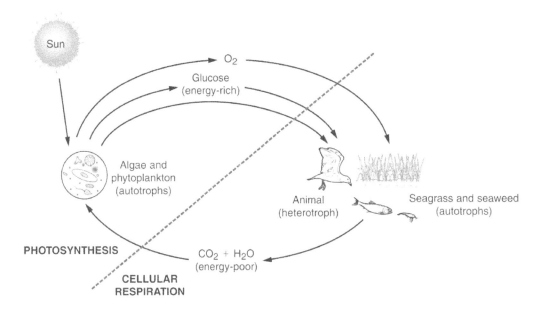

Figure 16-1 Summary of Photosynthesis and Cellular Respiration

The term dead zone was coined when fishers observed a fish kill, or the inability to catch shrimp or **demersal** fish in formerly productive fishing grounds. This was related to subsurface water that just did not have enough oxygen present to support many organisms. Different species have different oxygen requirements, and changes in the spatial distribution of organisms begin to appear as oxygen concentrations drop to 4.5 or 4 mg/L. Animals that are mobile will move. This affects the trophic structure of an area. Animals that cannot move are physiologically stressed. This may result in decreased growth or increased susceptibility to disease. If the oxygen levels keep dropping many organisms that are left die because they are unable to carry out aerobic cellular respiration. Cellular respiration and photosynthesis in marine systems (Figure 16-1) are only part of the story, with bacterial decomposition of excess algal biomass that has accumulated due to eutrophication having the potential to unbalance the entire cycle when hypoxic or anoxic conditions form and prevent many marine heterotrophs from surviving (Figure 16-2).

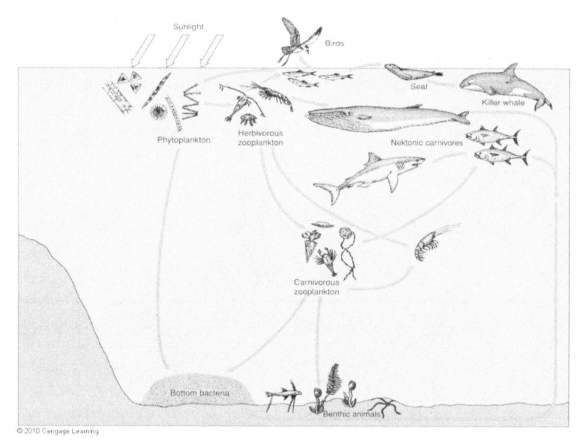

Figure 16-2 Accumulation of Excess Biomass Followed by Bacterial Decomposition Can Become a Dominant Factor in Eutrophic Systems

In mid spring to summer the continental shelf in the northern Gulf of Mexico experiences the second largest dead zone in the world (22,000 km^2 in 2002). This is the area that receives the outflow from the Mississippi and the Atchafalya Rivers (Figure 16-3). Two processes act together to result in zone of hypoxic water forming on the bottom of the shelf. First the water entering the gulf from the river basins is much less dense because it is warm and fresh, so the water column stratifies preventing surface waters that may have more dissolved oxygen from mixing with bottom waters. Second, cultural eutrophication occurs because of the very high nutrient levels in the river outflows. The dead zone is most intense May through August and then begins to break down as winter storms bring winds that mix the surface layer and disrupt the stratification (Rabalais et al. 2002). Dead zones do not mean that nothing living exists in the area, because aerobic organisms can still live at the surface and anaerobic organisms can live at the bottom, but the community structure is significantly changed.

This and other hypoxic zones in US coastal waters are of significant concern. Programs are in place to find ways to reduce nutrient run-off from and to enhance nutrient removal through natural processes. For example, efforts are being made to restore oyster populations as well as seagrass and marsh habitats in Chesapeake Bay (Committee on Environment and Natural Resources. 2010). Intensive research is being carried out to document the complex ecology involved well enough to put regulations and processes in place that will help solve the problem.

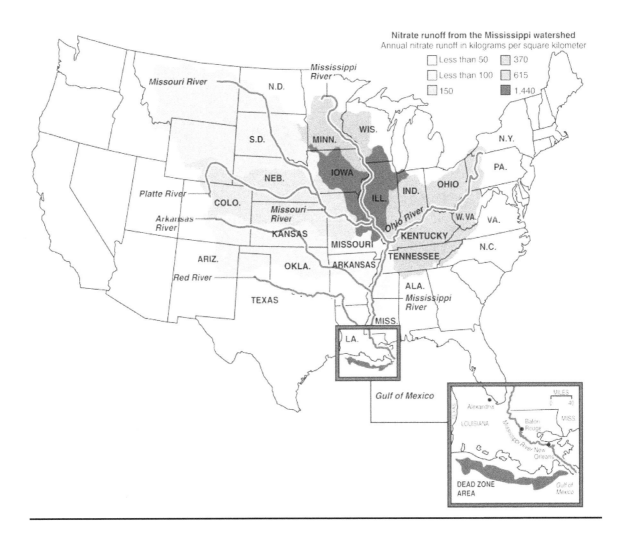

Figure 16-3 Mississippi Watershed and the Dead Zone in the Gulf of Mexico

Materials

- 4 L of seawater per group (or pond water*)
- Four 1-L clear soda bottles
- Liquid plant fertilizer, mixed according to manufacturer's directions
- (e.g., Miracle-Gro® All purpose liquid plant food, NPK ratio 15:30:15)
- GREEN Low Cost Estuary and Marine Monitoring Kit (for measuring pH, temperature, dissolved oxygen, turbidity, nitrate and phosphate, your lab may be equipped with alternative equipment, follow instructor directions)
- Pipettes for measuring fertilizer
- Graduated cylinders for measuring water
- Access to a window with uniform light exposure for all groups or Grow Lights
- Grease pencils or scientific tape and permanent markers for labeling
- (* if you are using pond water, use the GREEN Low Cost Water Monitoring Kit instead of the Kit designed for salt water)

Exercise 1
The Effect of Fertilizer on Water Quality

You will need to monitor the treatment groups in this exercise over the course of 2-3 weeks depending on the temperature of your lab and the initial concentrations of phytoplankton in your seawater.

1. Get into groups of 3-4 students.

2. Collect 4 clean, clear plastic soda bottles. Rinse each with approximately 50 mL of sea water. Label all bottles with the date and your group name.

3. Then label the bottles as Control, Treatment 1, Treatment 2 and Treatment 3.

4. Using the graduated cylinders provided measure out 750 mL of sea water into each bottle

5. Using the pipettes provided measure out 5 mL of liquid fertilizer into the bottle labeled Treatment 1, 15 mL into Treatment 2 and 25 mL into Treatment 3. Mix well after each addition of fertilizer.

6. Follow the directions provided with the water quality monitoring kit (or those for provided by your instructor for the equipment in your lab) and measure initial pH, temperature, dissolved oxygen, **turbidity**, **nitrate** and **phosphate** levels for all 4 treatment groups. Record data in Table 16-1.

6. Place bottles in window or under grow lights as directed by your instructor.

Following lab period(s):

1. Remove a water sample from each of your treatment groups according to the kit or your instructor's directions, and measure pH, temperature, turbidity, nitrate and phosphate levels. Make sure you are using clean glassware so you do not contaminate your samples. Record the data in Table 16-1.

Table 16-1 The Effect of Different Levels of Fertilizer on Water Quality.

Treatment Initial	pH	temperature	Dissolved oxygen	turbidity	nitrate	Phosphate
Control						
1						
2						
3						
Follow-up 1						
Control						
1						
2						
3						
Follow-up 2						
Control						
1						
2						
3						

a. What differences did you observe between your treatment groups?

b. Did the addition of the fertilizer result in hypoxia in all three experimental groups? Why or why not?

c. What processes important to the discussion of real hypoxic zones are NOT modeled in this experiment?

Exercise 2
Size of the Gulf Dead Zone over Time

You will use the data tracking the size of the dead zone in the Gulf of Mexico from 1985 to 2011 to look at trends. To better understand the variation in the data your instructor may also have you read the following article

Rabalais, N. N., Turner, R. E., Wiseman, W. J. (2002). Gulf of Mexico Hypoxia, A. K. A. "The Dead Zone." Annual Review Ecological Systems, vol. 33, pages 235–263.

1. Using the data provided in Table 16-2 plot the area of the dead zone in the Gulf of Mexico from 1985 to 2011. Put the year on the x-axis and the number of square kilometers on the y-axis.

2. Use the graph to answer the following questions:

   a. What is the trend in the size of the dead zone in the Gulf of Mexico over time?

   b. Why is the size of the dead zone so variable?

   c. 1993 was the year of a major flood even on the Mississippi River. What effect did that event had on the size of the dead zone? Explain.

d. Data were collected in midsummer over a 5 day period by a team of scientists on a research vessel. Research cruises in 2003, 2005, 2006, 2008 and 2010 were preceded by hurricanes. What effect do you think that had on the results?

e. What river outflow pattern would explain the data from 1988 and 2000?

f. Could the measurements of the **area** of the dead zone (km^2) be misleading if you were interested in changes in overall size?

Table 16-2 Gulf of Mexico Hypoxia. Area (km^2) of bottom water hypoxia. Data source www.gulfhypoxia.net. N. N. Rabalais, Louisiana Universities Marine Consortium, R. E. Turner, Louisiana State University; Funded by NOAA, Center for Sponsored Coastal Research.

Year	Square Kilometers	Year	Square Kilometers
1985	9774	1999	20000
1986	9432	2000	4400
1987	6688	2001	20720
1988	40	2002	22000
1989	no data collected	2003	8560
1990	9260	2004	15040
1991	11920	2005	11840
1992	10804	2006	17280
1993	17600	2007	20500
1994	16600	2008	20720
1995	18200	2009	8000
1996	17920	2010	20000
1997	15840	2011	17520
1998	12480	2012 forecast	16092

Questions

1. What characteristics of the Mississippi River system contribute to the formation of a dead zone?

2. Is the dead zone in the Gulf of Mexico always present? Why or why not?

3. How would weather affect the dead zone in the Gulf of Mexico?

4. What happens to the organisms in a dead zone?

5. Global climate change is predicted to increase rainfall in certain parts of the country and decreased rainfall in others and contribute to an increase in ocean temperatures. How would you expect this to affect the dead zone in the Gulf of Mexico? (hint check the predictions available in Bates et al. 2008 http://www.ipcc.ch/pdf/technical-papers/climate-change-water-en.pdf)

6. The Mississippi River watershed drains 41% of the contiguous United States, what would you suggest would need to happen to reduce the Dead Zone in the Gulf of Mexico? What are some of the challenges?

Suggested Reading

Committee on Environment and Natural Resources. 2010. Scientific Assessment of Hypoxia in U.S. Coastal Waters. Interagency Working Group on Harmful Algal Blooms, Hypoxia, and Human Health of the Joint Subcommittee on Ocean Science and Technology. Washington, DC.

Bates, B.C., Z.W. Kundzewicz, S. Wu and J.P. Palutikof, Eds., 2008: Climate Change and Water. Technical Paper of the Intergovernmental Panel on Climate Change, IPCC Secretariat, Geneva, 210 pp. http://www.ipcc.ch/pdf/technical-papers/climateschange-water-en.pdf

LUMCON (Louisiana Universities Marine Consortium) 2012. Hypoxia in the Northern Gulf of Mexico. www.gulfhypoxia.net.

Rabalais, N. N., Turner, R. E., Wiseman, W. J. 2002. Gulf of Mexico Hypoxia, A. K. A. "The Dead Zone." Annual Review Ecological Systems, 33:235–263.

Glossary of Terms

anoxic: Lacking oxygen e.g., water with no measureable levels of dissolved oxygen,

Atchafalya River: A short distributory of the Mississippi and Red Rivers that carries almost a third of the Mississippi's water into the Gulf of Mexico. The U.S. Army Corps of Engineers uses the Old River Control System to manage the flow of the Atchafalya, if this was not done it would become the main channel through which water flows into the Gulf of Mexico and the Mississippi would largely bypass Baton Rouge and New Orleans.

cellular respiration: A process completed by autotrophs, heterotrophs, and decomposers to convert organic compounds and oxygen into biologically usable energy (ATP), and release carbon dioxide.

cultural eutrophication: An increase in system productivity due to nutrient loading related to human activities.

demersal: Dwelling near or at the bottom of the water column.

dissolved oxygen: Measure of the amount of oxygen within water in mg/L (= ppm).

hypoxic: low levels of oxygen, e.g., water with low dissolved oxygen concentration (2 mg/L or less),

eutrophication: An increase in system productivity due to nutrient loading.

Gulf of Mexico: Body of water between Mexico and Florida, where scientists have found a seasonal hypoxic zone.

HAB: Harmful algal bloom, an explosion in the population of species of algae that release toxic chemicals.

Mississippi River: The major river in the central United States that empties into the Gulf of Mexico.

nitrate: NO_3^-, an ionic nitrogen compound that is important for plant and algal growth

Nonpoint source pollution: Pollution that comes from a diffuse source, e.g. agricultural or urban run-off.

phosphate: PO_4^{3-}, an ionic phosphorus compound that is important for plant and algal growth

Point-source pollution: Pollution with a single identifiable source

turbidity: A measure of water clarity

Lab 17

Why do Zoos and Aquariums Matter?

Objectives

Upon completion of this exercise, you should be able to:

1) Describe the importance of informal science education.

2) Become familiar with an informal science education facility (zoo, aquarium or science museum) and its resources.

3) Begin to assess the elements of exhibit design.

4) Describe a variety of career opportunities available in a zoo, aquarium or science museum.

Background: Informal Science Education

Zoos, science museums and aquariums play a very important role in our community. Over 200 million people a year visit a zoo or an aquarium every year in the United States. This represents a tremendous opportunity to engage the public with **informal science education**. In 1992 the American Association of Museums recognized education as being central to their role in public service and committed to making the educational experience more inclusive of diverse people.

Since that time a tremendous number of resources have gone into trying to understand how people of different ages, different levels of prior-knowledge, different cultural backgrounds and with different motivations learn in a **free-choice** environment like an aquarium. In addition to being places where research is done on the animals to learn about their basic biology or to enhance conservation programs, research is also done to assess visitor learning. The National Science Foundation (**NSF**), one of the major funding bodies in the county, funds projects focused on enhancing informal **STEM** (science, technology, engineering and math) learning. The America Association of Zoos and Aquariums (**AZA**), the Institute of Learning Innovation and the Monterey Bay Aquarium conducted a study to assess the impact of a visit to a zoo or an aquarium on the visitor's attitude towards conservation, stewardship and the love of animals. Entitled, "Why Zoos and Aquariums Matter", the study surveyed 5,500 people who visited 12 different institutions. The results indicated that zoos and aquariums had the ability to inform and to change attitudes about important environmental issues. Another study assessing public awareness about the ocean (conducted by AZA, The Ocean Project, Monterey Bay Aquarium, National Aquarium in Baltimore and **NOAA**) found that while, in general, people were initially not aware of ocean issues, a visit to an aquarium had the power to raise awareness, particularly among young people. Intensive market research is done by different groups to provide the data to strategically plan education and outreach programs.

It takes a tremendous number of trained professionals to help a zoo, a science museum or an aquarium fulfill its mission. Science students may want to explore the career opportunities available in the informal science education field. There are keepers and aquarists, but there are also curators, exhibit designers, exhibit technicians and education specialists of many kinds. Expertise in this kind of science education may become increasingly important as scientists try to find better ways to bridge the gap between environmental issues and public understanding. What works to make a visit meaningful to an individual in a free-choice learning environment? Not surprisingly since it is a complex question there are a set of complex answers and it is not the same for every person. Not only is it not the same for every person, it would not be the same for every person every time they went to their favorite museum for example. Many factors affect the experience e.g., motivation group size, how crowded the museum happens to be that day, and even something as simple as being hungry. The former Director of Visitor Research at the Exploratorium in San Francisco, Dr. Sue Allen writes about some of the goals for an effective exhibit (2004).

Immediate apprehendability or being able to look at something and quickly 'get it' is important in an environment where it is easy to just move on and where there are so many competing stimuli that 'overload' can happen quickly. It is easy to observe in a small crying child at the zoo-who has just 'had it' but it happens in adults too. There is tension between building an exhibit that is easy to understand quickly and one that is going to engage the observer more deeply by presenting some kind of challenge. **Physical interactivity** is another important component to consider. But not all learners are the same, so Allen suggests that having a **diversity of learning modes**, for example some quieter, more observational and reflective spaces are also important. Sometimes extra seating, changes in lighting or measures to reduce noise are all added to an exhibit space to help reduce physical and cognitive fatigue. It also helps to have good maps and other easily understandable tools to help a visitor orient. As a visitor moves from one exhibit to another are they able to pick up major themes? **Conceptual coherence** is important but a challenge when someone can wander through a series of displays or activities in a number of different ways of their choice (Allen 2004).

In this lab you will plan a trip to an informal science facility (individually, as a class or online), assess the effectiveness of the facility and design a field-trip assessment that would be appropriate for the mission of the facility, or your instructor may set aside a class period for Marine Science Day and have you develop and bring in an exhibit of your own.

Exercise 1
Orientation

People retain more of the details of an experience if they have oriented and have some idea of what to expect when they arrive at a new place. This exercise should be done ahead of time according to your instructor's directions)

Your instructor will tell you where this exercise will be carried out (if the class is going as a group) or you may get to choose a facility to visit. You can also do an online comparison of major aquarium resources (see Table 17-1).

1) Use the internet and see if the institution has an online presence. Use the website to answer the following questions:

   a) Where is the facility located? Are the directions clear? Do they include both driving and public transit options?

   b) What are the hours of operation? Are they open holidays? Do they have any special programs? Is entry free at any time?

   c) How much does it cost? Are there discounts? If there are discounts, for whom? Would cost be a barrier to some people?

   d) What resources are available ahead of time to an educator bringing a group? Can special resources be arranged ahead of time (e.g., a behind-the scenes tour, a sleep-over or access to a lunch room)?

   e) Does the facility host special events, e.g., talks or workshops? Will there be anything special going on the week you will be there?

f) Is there a map of the exhibits available online? What looks most interesting to you?

g) Does the facility have a jobs posting? Do they have student internships or volunteer opportunities?

h) What other resources were available on the website? Which ones were the most engaging or useful? What kind of information was not available that you felt would have been useful or relevant?

i) What kind of marketing tools does the institution use (e.g., print, TV and radio ads, social media)?

j) What is the institution's mission?

Exercise 2
Assessing Your Visit

As you visit the science museum, zoo or aquarium think about and record answers to the questions below. Use these questions as a starting place to assess the effectiveness of the facility. They are a guideline to help you think about different factors that may be relevant. You may think of additional criteria. Subjective terms such as 'adequate' and 'easy' are used. Your answers should include a description of how you defined these terms.

1) **Comfort Rating**

a) Was it easy to orient when you arrived?

b) Were the bathrooms easy to find? Well-maintained? Were they adequate for the size of the facility (i.e., was there a long line outside the woman's bathroom)? Was there a changing station for babies? Was there a companion restroom for a disabled person who needed assistance?

c) Was there adequate seating available for resting?

d) What facilities were available for food and drink? How were they priced? Could an average family of four afford a meal? Were the choices healthy and/or sustainable? If you had dietary restrictions were there options?

e) If you were pushing a stroller, on crutches or in a wheelchair, would the facility be easy to navigate (e.g., it is very difficult to use a walker on certain kinds of gravel paths)? If there were multiple floors, could you find the elevators?

f) Was an effort made to make the facility accessible to diverse groups of people in other ways, e.g., assisting a visitor who did not read English or someone who was not sighted? Look for elements of **universal design** e.g., a water fountain at a certain height might be accessible by someone who was younger and not very tall and by someone in a wheel chair.

g) Was the level of staffing adequate? Were the available **docents**/ volunteers helpful or approachable?

h) If appropriate, did the animals look healthy and well-cared for?

Other comments:

2) **Exhibit Rating**
    Visit and assess the exhibits in the facility. Your instructor might modify this depending on the details of the facility but a minimum of five exhibits is suggested.

    Immediate apprehendability

View the exhibits and decide if you get the point and then try to figure out what it was about the design that worked or did not work.

a) Would the exhibit 'work' at a variety of levels? e.g., was the signage appropriate for someone who did not have any science background as well as someone who did have prior knowledge? Did you learn something new?

Other comments:

3) **Physical Interactivity**
   a) Were there opportunities to interact physically in a meaningful way with any of the exhibits?

   b) Were any of the exhibits broken?

   c) Did the movement or the interactive feature get in the way of the content of the exhibit (e.g., a peek-a-boo exhibit where you can open a door and see picture of an animal where small visitors become intrigued by the sound the door makes when it opens and closes: playing with the door becomes the focus and not the animal behind the door)?

   d) Was the exhibit arranged in such a way that there was good access to the interactive exhibits? i.e., was traffic flow managed well?

Other comments:

4) **Diversity of Learning Modes**

    a) Did the exhibits engage a variety of sensory modalities e.g., seeing, hearing, or touching (tasting? Safety note: Do not lick the animals at the zoo)?

    b) Was there a balance between the exhibits suitable for different learning styles e.g., auditory (was there a guided tour on tape or a docent lecture), visual or kinesthetic (hands-on)?

    Other comments:

5) **Conceptual Coherence**

    a) Were the thematic connections between individual exhibits easy to figure out (e.g. in an aquarium or a zoo, organisms are often grouped by habitat)?

    b) Was the signage helpful and engaging or boring and/ or confusing? Was there any evidence visitors were actually reading the signs?

    c) Were there environmental issues being raised? Given your background knowledge on the subject, how effectively did you think the issue was being communicated?

    d) Was the exhibit organized so a visitor proceeded in a linear fashion to establish conceptual coherence or was there more 'free-choice'?

Other comments:

6) **Fun**

   a) Which exhibits seemed to be very popular, what characteristics did they share?

   b) Which ones did you like the best?

   c) What were the characteristics of the exhibits you found boring?

   Other comments:

Exercise 3
After Your Visit

Follow your instructor's directions as to which of the options will be required.

Option A
Write a report on your visit to a zoo, science museum or aquarium.

Use the answers to the questions in Exercises 1 and 2 and other notes to write an assessment. Answer the question: Did you think the facility accomplished its mission?

Option B
Design a rubric for assessing zoos, science museums and/or aquariums.

You now have some experience thinking about what makes an informal science experience effective. In the orientation section you identified the mission of the institution you were going to visit. Using criteria in exercise 2 that you think are relevant to the mission design a rubric to assess any similar institution. A rubric is like a grading guide. For example an instructor grading a research paper might use a grading rubric that looked something like the example below to evaluate a research hypothesis in a scientific paper. A full rubric would have a section for every significant part of the paper. Make your own table with a rubric that would work for 'grading' a zoo, science museum or aquarium.

Sample Rubric to Evaluate a Research Hypothesis in a Scientific Paper

	Excellent	Adequate	Poor
Hypothesis is supported by literature review in introduction	Use of resources logically supports hypothesis very well. Sources are current, extensive and cited properly	Use of sources supports hypothesis adequately. Sources may be dated, incomplete or there may be minor errors in citation format.	Hypothesis is not supported logically by sources. Sources inappropriately cited or missing entirely

Option C

Design an Exhibit for a Class Marine Biology Museum Day.

You now have some experience thinking about what makes an exhibit engaging and a good learning opportunity. Individually or in a group design an exhibit that you can bring to class and share. It can be interactive and physically engaging or observational and reflective.

Possibilities:

- A five minute video, highlighting a specific organism or marine environmental issue.

- A poster, highlighting a specific organism or marine environmental issue

- An interactive display or activity to illustrate a concept relevant to marine biology (e.g., tides or coastal erosion, or the way different gastropod shell types deal with compression stress as a protective mechanism.

- Use your imagination!

Questions

1. What job/career opportunities are available to someone who would like to work in a zoo, science museum or aquarium? Do some research to find out what kind of education would qualify someone for those positions?

2. Do you think zoos, science museums and aquariums matter? Why or why not?

Suggested Readings

Allen, S. 2004. Designs for learning: studying science museum exhibits that do more than entertain. Sci Ed. 88(Suppl.1):S17-S33. Published online in Wiley InterScience (www.interscience.wiley.com)

CAISE, Center for advancement of informal science education. http://caise.insci.org/
Falk, J. and J. Balling. 1977. An investigation of the effect of field trips on science learning. Final report grant number SED77-18913. Washington, DC: National Science Foundation.

Falk, J.H. and L. Dierking. 2000. Learning from Museums: Visitor Experience and the Making of Meaning. Rowman and Littlefield. 272 p.

Falk, J.H.; Reinhard, E.M.; Vernon, C.L.; Bronnenkant, K.; Deans, N.L.; Heimlich, J.E., (2007). Why Zoos &
Aquariums Matter: Assessing the Impact of a Visit. Association of Zoos & Aquariums. Silver Spring, MD.

National Research Council. 2009. Learning Science in Informal environments: people, places and pursuits. Committee on learning science innformal environments. Philip Bell, B. Lewenstein, A. W. Shouse and M. A Feder, Editors. Board on Science Edcuation, Center for Education, Division of Behavioral and Social Sciences and Education. Washington, DC. The National Academies Press.

The Ocean Project. 2011. America and the Ocean: Annual update 2011. http://www.theoceanproject.org/MarketResearch

Glossary of Terms

AZA: Association of Zoos and Aquariums (www.aza.org)

informal science education: Any exposure to scientific content or concepts that happens outside an organized classroom setting e.g., nature documentaries, a visit to a zoo, a park ranger talk while camping or a family trip to the ocean.

free-choice learning: Learning that happens throughout a person's lifetime based on the activities and experiences they choose

docent: A volunteer who serves as a guide in a museum

NOAA: National Oceanographic and Atmospheric Association

NSF: National Science Foundation

STEM: Science, Technology, Engineering and Math

universal design: Design of products and environments to be usable by all people to the greatest extent possible, without the need for adaptation or specialized design

Table 17-1 Suggestions for Aquarium sites to visit

Audubon Aquarium of the Americas, New Orleans	http://www.auduboninstitute.org/visit/aquarium
Georgia Aquarium, Atlanta	http://www.georgiaaquarium.org/
Monterey Bay Aquarium	http://www.montereybayaquarium.org/
Moody Gardens, Galveston	http://www.moodygardens.com/
Mystic Aquarium & Institution for Exploration, Mystic	http://www.mysticaquarium.org/
National Aquarium, Baltimore	http://www.aqua.org/
North Carolina Aquarium, Fort Fisher	http://www.ncaquariums.com/fort-fisher
Oregon Coast Aquarium, Newport	http://aquarium.org/
Ripley's Aquarium of the Smokies, Gatlinburg	http://www.ripleyaquariums.com/gatlinburg/
SeaWorld, San Diego	http://seaworldparks.com/en/seaworld-sandiego/
Shedd Aquarium, Chicago	http://www.sheddaquarium.org/
Tennessee Aquarium, Chattanooga	http://www.tnaqua.com/Home.aspx
The Florida Aquarium, Tampa	http://www.flaquarium.org/
Vancouver Aquarium Marine Science Center, Vancouver, B. C.	http://www.vanaqua.org/

Lab 18

Plastics in the Ocean

Objectives

Upon completion of this exercise, you should be able to:

1) Discuss the relationship between the density of different types of plastic and buoyancy in seawater.

2) Extract **microplastics** from household body care products and calibrate an **ocular micrometer** to determine the size of the particles.

3) Investigate the importance of sampling technique by using two different methods to recover plastic debris.

4) Discuss the complexity of the environmental issues related to plastics in the ocean.

Plastics in the Ocean

There are many important marine environmental issues such as global climate change, ocean acidification and overfishing. Plastic pollution in the ocean is yet another problem that has received a great deal of press in recent times. Scenes of birds starving to death with bellies full of plastic, turtles eating plastic bags and beaches covered with plastic debris are very much a part of the public consciousness. In this lab you will explore some of the issues that need to be considered when trying to quantify 'how much plastic is in the ocean' and 'what happens to the plastic in the ocean'.

Materials

- 1 beaker 500 mL
- 300 mL seawater (35 $^0/_{00}$, 20°C)
- Selection of 1 cm plastic samples from each of the 7 categories (small plastic caps, small pieces of pipe etc)
- Forceps
- Thermometer
- Refractometer

- 20 g Clean & Clear Daily pore cleanser (Johnson & Johnson)
- Microscope slides and coverslips
- 1 transfer pipette
- Stage micrometer
- Compound Microscope with ocular micrometer
- Three 250 mL beakers
- Stir bar
- Hot plate/stirplate
- Distilled water (200 mL)
- Hotpads
- Goggles
- 7.5 cm aquarium dip net with coarse mesh (approx. 2 mm)
- 7.5 cm aquarium dip net with fine mesh (approx. 100 μm)
- 15 g different types of plastic cut into pieces of different sizes
- 5 gal (19 L) rectangular aquarium filled with 15 L of seawater (35 $^0/_{00}$, 20 °C)
- Weigh boat
- Top-loading balance (g)
- Metric ruler or measuring tape
- Paper towels

Exercise 1
Does it Float?

Plastic is a generic word that can refer to a number of different hydrocarbon-based polymers. These polymers differ in their physical properties (Figure 18-1), which will affect their fate in a marine environment. How far a piece of plastic will travel in the ocean may be affected by whether or not it floats. Depending on its size a solid will float in a liquid if the density of the solid is less than the density of the liquid. If the density is very similar, the solid will be suspended in the liquid and if the density of the solid is greater than the liquid, the solid will sink. Density is the mass per unit volume of a substance. The average density of surface seawater is 1.027 g/cm^3. Since temperature, salinity and pressure affect density, density varies with location and season and is an important factor affecting vertical circulation and layered mixing in marine environments. Use Table 18-1 to predict the behavior of different kinds of plastic in seawater.

Table 18- 1 Properties of Plastic

Type of Plastic	Symbol and recycle code ♻	Density (g/cm^3)	Examples of types of items made with this type of plastic
Polypropylene	PP 5	0.90-0.92	Rope, diaper coverings, Tic-Tac lids
Low density polyethylene	LDPE 4	0.90-0.93	Milk jugs, packaging film, six-pack rings
High density polyethylene	HDPE 2	0.94-0.96	Bottle caps, garden furniture
Polystyrene	PS 6	1.03-1.06	Disposable cups, packing material
Polyvinyl Chloride	PVC 3	1.16-1.38	Piping, clothing
Polycarbonate	PC 7*	1.20-1.22	Eye glass lenses, toys, bottles
Polyethylene terephthalate	PETE 1	1.35-1.38	Polar fleece, bottles

* Not all plastics labeled 7 are polycarbonates.

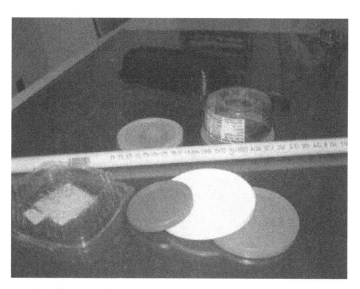

Figure 18-1 Typical Household Plastics of Resin Type 1 to 7 (Photo by author)

Table 18-2 Behavior of Plastic in Seawater

Type of Plastic (recycling number)	Prediction Float/suspended/sink	Observation Float/suspended/sink
Polypropylene (5)		
Low density polyethylene (4)		
High density polyethylene (2)		
Polystyrene (6)		
Polyvinyl Chloride (3)		
Polycarbonate (7)		
Polyethylene terephthalate (1)		

Does it Float?

1. Get into groups of 3-4 students. Determine the density of the seawater sample indirectly using salinity and temperature. Determine salinity using a hand held salinity refractometers (See Figure 18-2). These devices provide a direct reading of the specific gravity and concentration (in parts per thousand) of salt in water. The model available in lab has automatic temperature compensation and will already be calibrated for you.

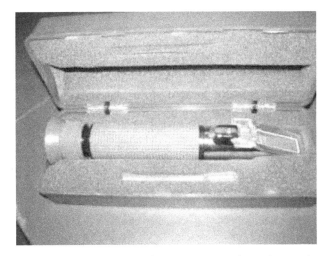

Figure 18-2 Salinity Refractometer (Photo by author)

2. Open the daylight plate and place 2-3 drops of the sample on the main prism of the refractometer.

3. Close the daylight plate so the sample spreads across the entire surface of the prism without air bubbles or dry spots.

4. Allow the sample to temperature adjust on the prism for approximately 30 seconds.

5. Hold the refractometer in the direction of a light source and look into the eyepiece. You should see a circular field with a graduated scale in the center. Take the reading where the boundary line of the blue and white cross the graduated scale. The scale will provide a direct reading of the concentration.

Sample salinity reading_____**‰**

6. Use the thermometer provided to measure the temperature in °C of the seawater sample.

Sample temperature reading_____**°C**

7. Assume the location of the lab is approximately sea level. If location of your institution is substantially different than sea level, your instructor will provide alternative instructions. Use the values determined in steps 5 and 6 and an online density calculator to verify the density of the sample of seawater you have been given (you can also compare it to the specific gravity reading).

Sample seawater density_____(g/cm^3)

8. Based on the density of the plastic given in Table 18-1 and the density of the seawater sample, develop hypotheses for the behavior the plastic items you are given and fill in your predictions for the behavior of each kind of plastic in Table 18-2.

9. Test your hypotheses by putting 300 mL of the seawater into a 500 mL beaker.

10. Choose a plastic item from a known recycling category. Pick up the plastic item with the forceps and gently place it in the beaker with the seawater. Submerge the item completely in the water, making sure that no air bubbles adhere to the surface of the item. Record the behavior of the item in Table 2.

11. Remove the item with forceps.

12. Repeat steps 10 and 11 for the remaining plastic items until you have observed the behavior of a piece of plastic in all the recycling categories.

    a) Were you able to predict the behavior of the plastic item in the seawater based on its recycling symbol?

    b) Based on your observations how do you think type of plastic will affect the fate of an item in the ocean?

Exercise 2
Microplastics

What are microplastics? Microplastics include any material that is solid, made of synthetic polymers and is 5 mm to .33 mm in size (Arthur et al. 2009). Small spheres of granulated polyethylene, polypropylene or polystyrene in that size range are now found in a wide range of body products (Gregory 1996, Fendall and Sewall 2009). These particles will find their way into the sewer system and eventually to the ocean as people brush their teeth or wash their face with products that now contain small plastic beads. The impact of these microplastics on the marine environment is not known. Graham and Thompson (2009) have shown that both filter feeding and deposit feeding invertebrates will ingest microplastics if they are present.   Scientists sponsored by government agencies, such as scientists in the NOAA Marine Debris program (http://marinedebris.noaa.gov/), are now asking questions such as, where are the microplastics in the environment (water column, sediments, organisms) and if present, are they causing any problems (http://www.tacoma.washington.edu/urbanwaters/research/microplastics.cfm)

1. Put 100 mL of distilled water into a 250 mL beaker and place on a hotplate to boil.

2. While the water is heating, place an empty 250 mL beaker on a balance, tare the balance to read zero, then squeeze approximately 20 g of the cleanser into the bottom of the beaker.

3. Remove the beaker with the cleanser from the balance and when the water in the other beaker is boiling, turn off the hotplate and, using appropriate safety precautions, pour approximately 50 mL of hot water onto the cleanser in the beaker.

4. Add a stir bar to the water and cleanser mixture and place the beaker on a stir-plate (make sure heating element is off) and stir until mixed thoroughly and much of the cleanser is dissolved.

5. Remove the cleanser solution and let settle for 3-5 minutes. You will see some sediment collecting on the bottom. Pour off the top 25-30 mL of the liquid portion, disturbing the sediment on the bottom as little as possible.

6. Being careful to draw the liquid from the bottom of the beaker, use a transfer pipette and make a wet mount of the remaining solution.

7. Observe what is on the slide using a compound microscope

8. Use a calibrated ocular micrometer to determine the size of the microplastic beads.

Calibrating an Ocular Micrometer

Measuring specimens present on a microscopic slide is possible by using a special tool mounted in one of the ocular lenses, called an **ocular micrometer**. An ocular micrometer consists of a glass disk on which fine lines have been etched. The distance between these lines must be determined and will be different depending upon the magnification of the objective lens that is in place. Determining that distance for each objective is called calibration. **Calibrating** an ocular micrometer requires a stage micrometer. A **stage micrometer** looks like a regular microscope slide, but on it are etched fine lines that are 0.01 mm (μm) apart (Figure 18-3). Use the stage micrometer to measure the distance between the lines on the ocular micrometer.

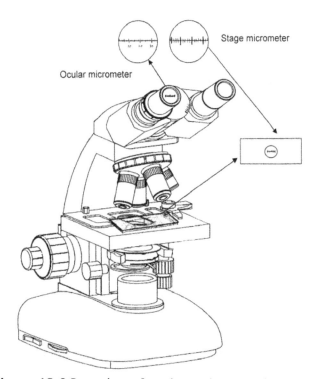

Figure 18-3 Location of ocular and stage micrometers

To Calibrate

1. Place the stage micrometer on the stage of the microscope and center the 'ruler' on the slide right under the light source.

2. Using the 4X objective lens, focus on the stage micrometer until you can clearly see the scale.

3. Adjust the relative positions of the ocular micrometer and the stage micrometer until the scales on both line up. You may have to rotate the ocular lens in order to do this.

4. Divide the number of stage micrometer divisions by the number of ocular scale divisions at the place where they line up exactly. You will not be able to line the entire length of the ocular micrometer scale with the entire length of the slide micrometer scale for all objective lenses. You may have to choose intermediate points-where the lines are precisely on top of one another (Figure 18-4). Multiply by 10 to convert to the metric unit of micrometers)

e.g., When using the 4X objective, if 16 stage units line up exactly with 26 ocular units (Figure 5) then

16/26 = 0.61 (therefore 1 ocular unit would be as wide as 0.61 stage units)

0.61 X 10 = 6.1 µm (Since each stage unit is 10 µm, 0.61 stage units would be 6.1µm wide)

Each ocular unit is 6.1 µm wide when using the 4X objective. You can now use the ocular micrometer to measure specimens on a slide.

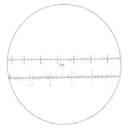

Figure 18-4 Stage and Ocular Micrometer

Student Questions:

1. What was the size of the microplastic beads found in the facial scrub?

2. Are these beads in the size range that would be taken in by filter feeding organisms?

3. Certain kinds of plastic preferentially adsorb toxins such as PCB's. Certain kinds of plastics also provide a substrate for attachment for bacteria and other microorganisms. What questions might a scientist ask about the effects of microplastics both positive and negative based on this information?

Exercise 3
How Much is Really There?

The ocean is a vast place. Trying to determine the amount of anything, e.g., fish population, phytoplankton, is difficult. Directly counting the number of anything is never possible, so scientists estimate, using a sample of a population and then extrapolating. Understanding the limitations of a sampling technique is critical to interpreting data appropriately. In this exercise you will use different sampling techniques to estimate the amount of plastic in a known volume of water (Figure 18-5).

Figure 18-5 The fate of different types of plastic in an aquarium tank containing seawater. (Photo by author)

1. Place weight boat on balance and record mass.

Mass of weigh boat _____**(g)**

2. Tare (zero the scale)
3. Measure out 15 g of plastic debris using weigh boat
4. Add 15 g of plastic debris to aquarium filled with sea water
5. Use forceps to submerge all plastic pieces and wait 5 minutes for pieces to float, suspend or sink.
6. Beginning at the short side of aquarium make one pass with the coarse-mesh dip net across the tank. Keep the top surface of the dip net at the surface do not let top part of the frame dip below the surface.
7. Remove contents of dip net, place on to paper towel to blot water
5. Place in weigh boat-using forceps to transfer any of the smaller pieces?
6. Determine mass of plastic recovered
7. Use the metric ruler or measuring tape measure the height and width of your sampling net. Measure the length of your sampling distance (i.e., side to side in the aquarium). Calculate the volume of water sampled.

H_____cm x **W**_____cm x **L**_____cm =**Volume of water sampled** _____cm^3

8. Use the metric ruler or measuring tape measure the height, width and length of the water in the aquarium. Calculate the volume of water in the tank.

H_____cm x **W**_____cm x **L**_____cm =**Volume of water in tank** _____cm^3

9. Repeat using fine-mesh aquarium dip net.
10. Record data in Table 3.

Table 18-3 Sampling for Plastics

	Coarse mesh net	Fine mesh net
Volume of water sampled (cm^3)		
Plastic recovered in sample (g)		
Estimate of plastic density in water sample (g/cm^3)		
Total volume of tank (cm^3)		
Known amount of plastic in tank (g)		
Known plastic density in tank (g/cm^3)		

a) How does the estimate of the plastic density compare to the known density? Compare using the two different types of nets?

b) Were there any other variables that affected how effective the sampling method was at estimating the amount of plastic in the system?

c) **Neuston** (surface-dwelling organisms) sampling nets are typically 333 μm mesh. How does this compare to the size of the microplastic beads found in the body wash. Do you think we have a good estimate of the plastics in our oceans?

d) What is the effect of sampling method on our estimate of the amount of plastic in the tank? Do you think this would be relevant in sampling marine environments? Why or why not?

e) Comment on sources of error in your sampling method.

Questions

1. Using the Internet, define the "Pacific Garbage Patch."

2. What are some of the known issues related to plastics in the marine environment?

3. Read the commentaries posted on the issue of the Pacific Garbage patch at the three sites listed below and answer the following questions.

   - Goldstein, M. 2011. Does the "Great Pacific Garbage Patch" even exist? << SEAPLEX. Retrieved on April 27, 2011 from http://seaplexscience.com/2011/01/10/does-the-great-pacific-garbage-patch-exist/

   - Hoshaw, L. 2009. Afloat in the Ocean, Expanding Islands of Trash. Retrieved on April 27, 2011 from http://www.nytimes.com/2009/11/10/science/10patch.html

   - White, A. 2011. Oceanic "garbage patch" not nearly as big as portrayed in media | Oregon State University. Retrieved on April 27, 2011 from http://oregonstate.edu/urm/ncs/archives/2011/jan/oceanic-%E2%80%9Cgarbage-patch%E2%80%9D-not-nearly-big-portrayed-media

a) Is there any controversy related to the size of the garbage patch? Explain.

b) Is there any controversy related to the importance of plastic pollution in the marine environment? Explain.

3. What are the options for recycling different kinds of plastic where you live?

4. Survey the body care products in your home or in the facial cleanser aisle of a grocery or drug store for the presence of microplastic beads. Look particularly at 'scrubs' or exfoliating cleansers. List the items that contain an ingredient called polyethylene (just polyethylene not polyethylene glycol) contain small spheres of plastic.

5. Metric system review:

       1 cm = _____ mm

       1 mm = _____ μm

Optional Activity:

If you have the ability, permission, and space, for a 1 week period, wash, dry, collect and save all the plastic material that you generate (e.g., See Figure 18-6) that cannot be recycled easily from your home. At the end of the week, bring the collected plastic into the lab so the mass of plastic generated can be determined.

Figure18-6 Household plastic waste (Photo by author)

Mass of plastic generated in 1 week _____ g x 52 weeks/year = _____ plastic per year.

The EPA (http://epa.gov/osw/nonhaz/municipal/msw99.htm) estimates 2 kilograms (about 4.5 lb) waste/person/day. Approximately 12% of this waste is plastic.

2 kg per person/day x 0.12 x 365 days = _____average amount of plastic per year

How do your results compare to annual plastic consumption in the US and can you think of ways to reduce this waste stream?

Suggested Readings

Arthur C, J Baker and H Barnford. Eds. 2009. Proceedings of the International Research Workshop on the Occurrence, Effects, and Fate of Microplastic Marine Debris, September 9-11, 2008. National Oceanic and Atmospheric Administration Technical memorandum NOS-OR&R-30. EPA, 2011. Municipal Solid Waste (MSW) in the United States: Facts and Figures. Retrieved May 20, 2011. From http://epa.gov/osw/nonhaz/municipal/msw99.htm

Boerger, CM, GL Lattin, SL Moore and CJ Moore. 2010. Plastic ingestion by planktivorous fishes in the North Pacific Central gyre. Marine Pollution Bulletin. 60: 2275-2278.

Fendall, LS and MA Sewell. 2009. Contributing to marine pollution by washing your face: microplastics in facial cleansers. 58: 1225-1228.

Graham, ER and JT Thompson. 2009. Deposit-and suspension-feeding sea cucumbers (Echinodermata) ingest plastic fragments. J. of Exp. Marine Biol and Ecology 368: 22-29.
Gregory, M. 1996. Plastic 'scrubbers in hand cleansers: a further (and minor) source for marine pollution identified. Marine Pollution Bulletin. 32: 867-871.

UW Center for Urban Waters. 2011. Sources and distribution of marine microplastics. Retrieved April 27, 2011 from http://www.tacoma.washington.edu/urbanwaters/research/microplastics.cfm

Goldstein, M. 2011. "Does the "Great Pacific Garbage Patch" even exist?" SEAPLEX. Retrieved on April 27, 2011 from http://seaplexscience.com/2011/01/10/does- the-great-pacific-garbage-patch-exist/

Hoshaw, L. 2009. "A Float in the Ocean, Expanding Islands of Trash." Retrieved on April 27, 2011 from http://www.nytimes.com/2009/11/10/science/10patch.html

Law, KL, S Moret-Ferguson, NA Maximenko, G Proskurowski, EE Peacokc, J Hafner, and CM Reddy. 2010. Plastic Accumulation in the North Atlantic Subtropical Gyre. Science. 329: 1185-1188.

Masura, J. "Marine debris-fishing for microplastics in your home." Retrieved on April 27, 2011 from http://e3washington.org/upload/profile/resources/file-227.pdf

NOAA. 2011. Marine Debris. Retrieved April 27, 2011 from http://marinedebris.noaa.gov/
Tomaczak, M. 2000. Seawater Density Calculator. Retrieved on April 27, 2011 from http://www.es.flinders.edu.au/~mattom/Utilities/density.html

Rios, LM, C Moore, PR Jones. 2007. Persistent organic pollutants carried by synthetic polymers in the ocean environment. Marine Pollution Bulletin 54: 1230-1237.

Ryan, PG, CM Moore, JA van Franker and CL Moloney. 2009. Monitoring the abundance of plastic debris in the marine environment. Phil. Trans. R. Soc. B 364:1999-2012

Suthers, IM and D. Rissik, eds. 2009. Plankton, A guide to their ecology and monitoring for water quality. CSIRO

White, A. 2011. Oceanic "garbage patch" not nearly as big as portrayed in media Oregon State University. Retrieved on April 27, 2011 from http://oregonstate.edu/urm/ncs/archives/2011/jan/oceanic-%E2%80%9Cgarbage-patch%E2%80%9D-not-nearly-big-portrayed-media

Acknowledgements:

Linda Fergusson-Kolmes would like to thank Dr. Jan Hodder and Coral Gerke, University of Oregon and COSEE Pacific Partnerships for providing the funding to support the Community College Faculty Institute 2010: Recent Advances in Oceanography. Dr. Angelique White, Oregon State University for her presentation and discussion of the issue of plastics in the marine environment at the COSEE summer workshop 2010. Julie Masura University of Washington at Tacoma, for the demonstration of microplastic sampling techniques and sharing her lab ideas.

Glossary of Terms

Microplastics: Material that is solid, made of synthetic polymers and is 5 mm to .33 mm in size

Neuston: Surface-dwelling organisms

Ocular micrometer: Measuring device inserted into the eyepiece of a compound microscope to allow measurements to be taken of specimens on a slide

CPSIA information can be obtained
at www.ICGtesting.com
Printed in the USA
BVHW012031220922
647701BV00012B/54